What Every Engineer Should Know About

Threaded Fasteners

Materials and Design

WHAT EVERY ENGINEER SHOULD KNOW
A Series

Editor

William H. Middendorf

Department of Electrical and Computer Engineering
University of Cincinnati
Cincinnati, Ohio

Other volumes in preparation

What Every Engineer Should Know About

Threaded Fasteners

Materials and Design

Alexander Blake

Lawrence Livermore National Laboratory
University of California
Livermore, California

CRC Press

Taylor & Francis Group
Boca Raton London New York

CRC Press is an imprint of the
Taylor & Francis Group, an **informa** business
A TAYLOR & FRANCIS BOOK

First published 1986 by Taylor & Francis

Published 2019 by CRC Press
Taylor & Francis Group
6000 Broken Sound Parkway NW, Suite 300
Boca Raton, FL 33487-2742

© 1986 by Taylor & Francis Group, LLC
CRC Press is an imprint of Taylor & Francis Group, an Informa business

First issued in paperback 2019

No claim to original U.S. Government works

ISBN 13: 978-0-367-45157-8 (pbk)
ISBN 13: 978-0-8247-7554-4 (hbk)

Visit the Taylor & Francis Web site at
http://www.taylorandfrancis.com

and the CRC Press Web site at
http://www.crcpress.com

Library of Congress Cataloging-in-Publication Data

Blake, Alexander.
   Threaded fasteners.
      p. cm. (What every engineer should know ; vol. 18)
   Includes bibliographical references and index.
   ISBN 0-8247-7554-6
   1. Fasteners. 2. Screw-threads. I. Title. II. Series: what every engineer should know ;v. 18.
TJ1338.B57 1986
621.8'82   86-13527

Library of Congress Card Number 86-13527

Preface

This reference book is designed to serve the needs of engineers, technicians, designers, and technologists concerned with standards, materials, and elementary formulas for the selection, procurement, and quality control of fasteners. The intent is to dispel some of the puzzles and to minimize distinct frustration over the bewildering maze of fastener options. No special knowledge of mechanics or metallurgy is required since the material consists of basic definitions, properties, and examples of calculations suitable for design, manufacturing, or sales offices. In addition, this book should be of use to students and teachers in technical schools establishing design and industry-oriented curricula.

The presentation is set around the considerable body of materials and standards knowledge developed by Richard Belford, who for many years served as Managing Director of the Industrial Fasteners Institute in Cleveland, Ohio. Belford's distinguished career in promoting both national and international standards is second to none. If even in a modest way, this volume reflects such contributions, then the time and effort of the reader interested in fastener technology will indeed be well spent.

Alexander Blake

Contents

Contents

About the Author

ALEXANDER BLAKE is Engineer-at-Large at the Lawrence Livermore National Laboratory, Livermore, California. In this capacity, he presides over engineering analysis, mechanical design, standard practices, quality assurance, and safety committees. The author of numerous technical reports, papers, and articles on engineering design, stress analysis, and solid mechanics, he has written three books, including *Practical Stress Analysis in Engineering Design* and *Design of Mechanical Joints* (both titles, Marcel Dekker, Inc.). He is also Editor-in-Chief of *Handbook of Mechanics, Materials, and Structures*. Mr. Blake was awarded the title of Chartered Engineer by the British Council of Engineering Institutions in 1969 and was elected a Fellow of the American Society of Mechanical Engineers in 1974. He is listed in *Who's Who in the West, Who's Who in Engineering*, and *American Men and Women of Science*. Mr. Blake received his graduate and postgraduate training in Mechanical Engineering from the University of London in England.

1

Introduction

Volumes can be written about the development of mechanical fasteners, which date back to Herculaneum in 79 A.D. The use of a bolted joint, however, became truly practical during the 18th century. Since then much valuable information on fastener design, testing, manufacture, and service has been developed in such countries as the United States, United Kingdom, Germany, Holland, Austria, Japan, France, Belgium, Switzerland, and Italy. The time interval between the Second World War and the close of the 1960s was most productive and it might be termed the "golden era" of fastener development.

About half a century of progress in developing fastener standards in the United States has centered around the procedures of the American National Standards Institute, Inc. (ANSI) and the Industrial Fasteners Institute (IFI). The design engineers and managers of manufacturing and construction industries can consult ANSI and IFI standards covering dimensions, geometry, and practice for a great number of mechanical fasteners. Further detailed information is also available from engineering and production handbooks. This volume provides a guide to the dimensional, geometrical, and materials aspects of the fastener, together with fundamental design formulas, as a service to practical engineering designers faced with technical and economical decisions in relation to the performance of mechanical connections.

The following additional acronyms are used throughout this book:

AISI   American Iron and Steel Institute
ASTM   American Society for Testing and Materials
CDA   Copper Development Association
ISO   International Organization for Standardization
SAE   Society of Automotive Engineers
UNS   Unified Numbering System (for Metals and Alloys)

Although the mechanics of a bolted joint, for instance, is not a mundane engineering topic there are several essential features of geometry, materials, and design of fasteners which the generalist can appreciate without being an experienced metallurgist, designer, or stress analyst. This book assumes little specialized knowledge in the field and it is intended to serve the practical needs of a design office, test laboratory, and production department. The material covers the areas of definitions, dimensions, thread forms, strength properties, fastener grades, materials, and elementary design formulas sufficient to make a procurement request or a preliminary design estimate. The area of materials selection is particularly important for design purposes under various environmental conditions. For this reason the book includes many practical details of material properties, composition, fabrication, and classification. The attempt is made to offer a complete package of information that an engineer should have for his every day encounters with fasteners.

The name *bolts* is often extended to all threaded fasteners such as studs and machine screws, and the number of different fasteners defined by the various standards in the world is truly enormous. This situation, combined with the long and tedious process of conversion of all specifications to one international standard, creates additional difficulties in correlating information on the theory and practice of fastener behavior.

This problem is complicated further by the fact that the methods of dealing with fastener behavior are of a statistical nature. The larger the total population of fasteners and joints we are trying to understand the more data appears to be needed to get a reasonable prediction of fastener performance. In the real world, however, we can only afford to test a few samples. We are, of course, interested in the scatter of test results and the more variation we find, the less certainty we have about our estimates. The grouping of test results may or may not follow the so-called normal distribution diagram. Such diagrams can give us the overall picture of fastener behavior in which it may be difficult to zero in on a particular variable. If we can have a relatively homogeneous population of the properties tested, then 25 test samples should provide a reliable result.

Although many tests have been documented for individual threaded fasteners, and considerable progress has been made in numerical analysis, our knowledge of fastener behavior continues to be incomplete. Reliance, therefore, has to be based on our own experience, engineering handbooks, and regulatory documents. The areas of new material properties and the quality of fabrication are particularly difficult to define. The specifications and standards developed in many countries are intended to reduce the cost of this process, and to promote reliability.

Several typical examples of bolt configurations are shown in Fig. 1. The design of these fasteners should follow a number of standard procedures such as, for instance, the following:

Determine the external force for the most highly loaded fastener
Select a factor of safety based on the type of application and best engineering judgment
Select a fastener material for given environmental and economic constraints
Select fastener dimensions
Calculate the required tightening torque

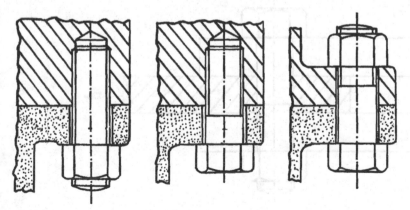

Figure 1 Typical bolt configuration.

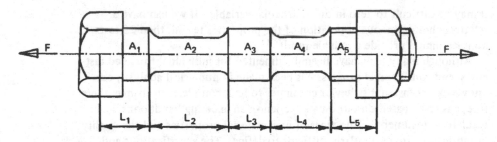

Figure 2 Example of a complex shape fastener.

The fasteners shown in Fig. 1 have uniform cross-sections. The problem of design and application however can be somewhat different when the geometry of the shank varies as indicated in Fig. 2. It is clear from Fig. 2 that in addition to the changes in fastener diameter, the transition radii may become of importance in selecting the maximum allowable axial load on the fastener.

The condition of shear loading on a particular fastener depends on the application and the number of shear planes involved. The cross-section in the case of a threaded fastener can be either through the main body of the

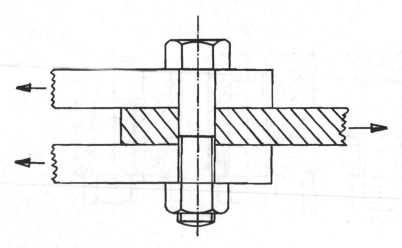

Figure 3 Example of fastener in double shear.

shank or the threaded region. The application shown in Fig. 3 involves both types of cross-sections.

While our primary interest may be tensile, rather than shear strength of the fastener it is important to recognize that the tensile strength may be reduced when torsional or shear loads are present. Such effects, however, are best determined experimentally.

2

Definitions of Screw Threads

Screw threads are remarkably complex machine elements; for that is exactly what they are. When properly assembled they have a capability to develop and transfer tremendous forces. There are over 125 separate geometrical features and dimensional characteristics in the design and construction of screw threads, each with its own term and definition. However, with a familiarity of only about 30 the engineer can be comfortably conversant in the language of screw threads and, more importantly, can have a good working understanding of their function and performance capabilities.

A *screw thread* is a ridge of uniform section in the form of a helix on the external or internal surface of a cylinder. *External threads* are screw threads on bolts, screws and studs; *internal threads* are in nuts and tapped holes.

The configuration of the thread in an axial plane is its *profile* and the three parts making up the profile are the *crest*, *root*, and *flanks* (see Fig. 4). The crests of threads are at the top, the roots at the bottom and the flanks join them. A thread having full form at both crests and roots is a *complete* (or *full form*) *thread*. When either crest or root is not fully formed it is an *incomplete thread*. Incomplete threads occur at the ends of externally threaded products which are pointed, at the runout where the threads

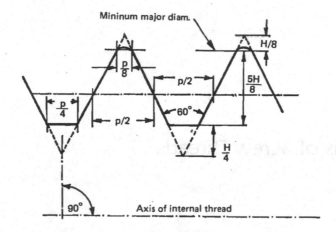

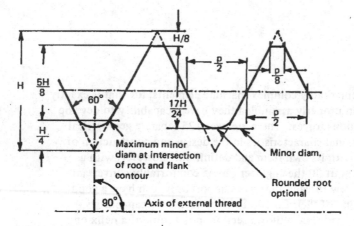

Figure 4   Unified thread form.

blend into an unthreaded shank, or at the countersinks in the faces of nuts and tapped holes.

On an external thread, the diameter at the thread crest is the *major diameter*, and at the thread roots is the *minor diameter*. On an internal thread it is just the opposite, the diameter at the thread crest is the *minor diameter* and that at the thread root is the *major diameter*. The angle between a flank and a perpendicular to the thread axis is the *flank angle*. When both flanks have the same angle we have a *symmetrical thread* with the flank angle termed *half-angle of thread*. Unified and ISO metric threads have a 30° flank angle and are commonly known as 60° threads because this is the included angle between their thread flanks.

The *fundamental triangle height* H is the height of the thread when the profile is extended to a sharp V form. For a 60° thread H equals 0.866025 times the thread pitch. *Thread height* is the distance measured radially between the major and minor diameters. In American practice thread height is usually expressed as a percentage of 0.75H. Several design details of unified screw threads are given in Fig. 4.

Thread pitch (p) is the distance measured parallel to the thread axis between corresponding points on adjacent threads. Unified threads are designated in *threads per inch*, consequently their thread pitch is the reciprocal of the number of threads per inch. Metric threads, however, are designated using actual thread pitch. *Thread series* are groups of diameter-pitch combinations distinguishing one from the other by the number of threads per inch or thread pitch applied to a series of specific diameters. For fasteners in general the most popular inch thread series are coarse and fine. Often used metric series involves coarse thread.

Pitch diameter is the diameter of a theoretical cylinder that passes through the threads in such a position that the widths of the thread ridges and thread grooves are equal. On a perfect thread these widths should each equal one-half of the thread pitch. The *thread axis* is the axis of the pitch cylinder. With anything less than a perfect thread, the actual pitch diameter, as measured at any position throughout the length or circumference of the thread, will vary depending on the variations in manufactured thread form within the permitted limits of size. Consequently, the definitions of measurement and significance of pitch diameter are quite controversial among screw thread experts. At the same time these values are important for computation and thread design purposes, production of manufacturing tools and dies as well as thread acceptance gages and gaging. The *pitch line* is the generator of the pitch cylinder.

The *basic thread profile* establishes an absolute boundary between the product external and internal threads. If either trespasses beyond this boundary then potential interference exists and product threads may not assemble. It is from this basic thread profile that the actual *limits of size* are derived by applying allowances and tolerances. An *allowance* is an intentional clearance between the mating threads. In other words, when both the external and internal threads are manufactured to their absolute *maximum material condition* there will be a finite space between them. For fasteners the allowance is generally applied to the external thread which means that its maximum diameters are less than basic by the amount of the allowance. The minimum diameters of internal threads (its maximum material condition) are basic. *Tolerances* are specified amounts by which dimensions are permitted to vary for manufacturing convenience. The tolerance is the difference between the maximum and minimum permitted limits. Thus, for external threads, the maximum material condition less the tolerances (moving toward the thread axis) defines the *minimum material condition*. For internal threads, the maximum material condition plus the relevant tolerances (moving away from the thread axis) define the minimum material condition.

The combination of allowances and tolerances in mating threads is *fit* which is a measure of tightness or looseness between them. A *clearance fit* is one that always provides a free running assembly. On the other hand an *interference fit* is characterized by specified limits of thread size which ensure a positive interference between the threads when assembled.

When assembling externally threaded fasteners into internally threaded nuts or tapped holes the axial distance through which the full threads of each are in contact is *length of thread engagement*. The distance the mating threads overlap in a radial direction is *depth of thread engagement*. It is determined by the major diameter of the external thread minus the minor diameter of the internal thread.

Strengths of screw threads, that is their ability to support and transfer loads, are dependent on four stress areas. However there are five possible thread failure modes, three of which are critical and two that are not too common. The externally threaded part can fracture in tension through its threads, the external thread can strip off, and the internal thread can strip out. The other two modes are shear through the threaded section normal to the thread axis or torsion, and twist-off due to overtightening. *Tensile stress area* is the assumed area through the thread used to compute the tensile load carrying capacity of fasteners. The tensile stress area is equivalent to the cross-sectional area of a theoretical cylinder of the same material and mechanical properties capable of supporting the same ultimate tensile load. *Thread shear area* is the effective area through the thread ridges, parallel to the thread axis and spread over the full length of the thread engagement supporting the applied load in shear. The shear plane for internal threads occurs at the major diameter of the external thread, and for external threads at the minor diameter of the internal thread. *Thread root area* is the cross-sectional area through the external thread of its minor diameter. This area is used to evaluate the susceptibility of the thread to either a cross-sectional or torsional shear failure.

3

Thread Forms

There are literally dozens of different screw thread forms. However, for conventional mechanical fasteners only the following five forms have significance: UN, UNR, UNJ, M, and MJ. All are defined as 60° threads of essentially the same basic profile. The principal differences between the various forms can be traced to the contour of the root of the external thread.

Prior to 1948 the American National thread form (Fig. 5) was the standard in North America. In that year, the United States, Canada, and Great Britain agreed to adopt a single screw thread system to supplant the American National in the United States and Canada, and the Whitworth in Great Britain. The new system was called *Unified* and it is currently recognized as the standard throughout the world for all inch series fasteners. In a metric type screw thread (Fig. 6) the radius is expressed in terms of the triangle height H.

The Unified thread form—designated UN—is practically identical with the now obsolete American National. In fact, threads manufactured to either of the standards are functionally interchangeable. The UN thread form, as originally designed, provided for either flat or rounded roots in the external thread at the option of each country's national standards. The United States chose to permit flat roots, even though there was a complete understanding that stress concentrations in threads could be lowered by rounding their

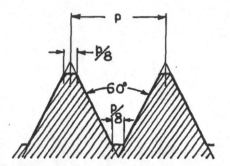

Figure 5 American standard screw thread.

roots. However, economics dictated. Thread-roll dies and thread cutting
tools at that time were expensive, particularly in fabrication of the radiused
crests. Also influencing this decision was the production fact-of-life that new
tools wear so that within the manufacture of few hundred parts the thread
crests of the tool becomes rounded off. Consequently the roots in the product thread no longer remain sharp, but take on a contoured form.

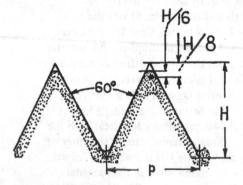

Figure 6 International screw thread.

As performance demands increased, particularly for safety-critical fasteners subjected to fatigue-inducing loads, it became imperative to search out opportunities for fatigue behavior improvement. An obvious answer was to no longer depend on tool wear but to actually require controlled radiusing of the external thread root. This led to the design and introduction of a modified thread form, designated UNR, with the single difference from UN being a mandatory root radius having the limits of 0.108 to 0.144 times the thread pitch. The minimum radius of 0.108p is the largest radius that can be fitted into the UN profile without violating the minimum material condition of the external thread, and 0.144p is the largest radius that can be accomodated without causing theoretical interference with an internal thread at its maximum material condition. The UN thread completely encompasses the UNR form. Again, the only difference is that UNR external threads must have a radiused root.

When first introduced in the late 1950s it became necessary to specify the UNR form to assure delivery of fasteners with radiused roots. However, today virtually 100% of all inch series fasteners in nominal sizes 1 in. and smaller are manufactured with radiused roots whether UNR is specified or not. This is because threads in this size range are today normally produced by thread rolling utilizing thread roll dies with standard rounded crests. For larger size fasteners, which are beyond the capability of thread-rolling equipment and whose threads are cut, UNR should be specified whenever the radiused form is needed. Otherwise the manufacturer will probably supply the UN form of the fastener.

Even before UNR threads were introduced, extensive research was in progress to develop an optimum thread form, one that would have superior fatigue resistance without sacrificing the static strength. In other words the question was just how generous could the root radius be to satisfy the above criteria. The answer was a new thread form, designated as UNJ. UNJ threads are basically UNR thread forms with a larger mandatory root radius having the limits of 0.150 to 0.180 times the thread pitch. Because of the enlarged radiusing it became necessary to modify the internal thread by increasing its minor diameter in order to prevent any possibility of interference with the external thread root during assembly. UNJ threads are now standard for all fasteners used in aerospace and many other critical applications.

UN internal threads assemble correctly with UN and UNR external threads. Theoretically, UN internal threads will not assemble with UNJ external threads. However, this combination has been used by major fastener users for several years now claiming trouble-free experience. Computer studies also substantiate that the statistical risk of actual interference between the manufactured parts is negligible. Even so, this practice is not recommended especially if the fasteners are plated. UNJ internal threads assemble correctly with UNJ external threads and also fit both UN and UNR. However, this latter mating should be used with a degree of caution because the increased minor diameter of the UNJ internal thread reduces the stripping area of the external thread.

In 1969, ISO 68 was published establishing for the first time the basic thread profile of an international screw thread system. This profile, which is now commonly designated M, has exactly the same geometry as the UN thread profile, and like the UN form it permits a flat root in the external thread with a special recommendation of a minimum radius of 0.1 times thread pitch. During the follow-up ISO Meetings North American spokesmen argued in favor of adopting as the world standard for metric commercial fastener threads a metric converted UNJ thread form. Delegates from other countries, while agreeing with the potential of beneficial effects of root radiusing, were reluctant to accept the enlarged UNJ roots because of the accompanying need to alter the internal threads as originally defined in ISO 68. The eventual compromise was that M profile threads would have a mandatory rounded root in their external thread with a minimum radius of 0.125 times thread pitch for all fasteners having specified tensile strengths of 800 MPa (116 ksi) and greater. For lower strength fasteners the root would preferably be rounded but it could also remain flat. In the United States virtually all metric fasteners in nominal thread diameters M24 and smaller are roll-threaded and consequently have radiused roots regardless of their strength properties. Unlike UN however, for sizes larger than M24 all higher strength fasteners must have radiused roots.

As metric interest intensified in the United States it became apparent that there would be a need for a metric counterpart of the UNJ thread form. Such a thread, designated MJ, was introduced in North America in the early 1970s and has since been accepted in principle for adoption internationally as an ISO standard. MJ threads are identical in their geometry to UNJ form.

Thus there are just two metric screw thread forms, the M profile which is the standard for commercial fasteners and the MJ which is the standard for aerospace quality fasteners.

It is interesting to reflect on the five thread forms for fasteners just discussed and to observe that they are essentially the same with the one exception involving root contour in the external thread. In summary then, while in the case of UN threads there is no mandatory minimum root radius, UNR requires 0.108p, M profile has 0.125p, and UNJ and MJ forms recommend 0.150p. It is difficult to believe that such modest differences could be important, but they are.

Radiusing the root of external threads adds only slightly to the tensile strength of a fastener subjected to static tensile loading. And the reason is one of simple geometry. As the root radius increases minor diameter increases and the cross-sectional area through the thread becomes greater. However, the amount of area growth is so small that the same tensile stress area formula is used in stress calculations for all thread forms.

It is rare to find an assembly joined by mechanical fasteners which is not, to some degree, exposed to dynamic loading during its service life. Extremely few mechanical joints remain static and totally insulated from some form of fluctuating stress, vibration, stress reversal or impact. Fortunately, in only a small percentage of all joints are the fatigue resistant properties of the fastener itself the primary design consideration. When they are, no opportunity for improvement can be ignored. And it is here that root radiusing really counts. The larger the root radius, the better the fatigue resistance of the fastener.

Fatigue failures of stressed parts generally occur at locations of high stress concentration, such as notches or abrupt changes in cross-sectional configuration. Screw threads with their large variations in cross-section and their thread roots which act as notches are particularly susceptible to fatigue. The highest stress concentrations in threads occur at the roots and the magnitude of the stress concentration factor relates directly to whether the root is rounded and to what degree.

Computation of thread stress concentration factors is an exceptionally complex exercise and the answers are not always reliable. Consequently, physical research programs, including photoelastic studies, have been conducted to investigate the influence of root radiusing on fatigue properties of threaded sections. A generalized conclusion is that when all other variables, such as fastener size, thread pitch, material, manufacturing method, etc., are uniform with the only variable being root radius, stress concentration factors can be reduced from about 6 for sharp or flat roots to less than 3 for UNJ and MJ threads. Translated this means a possible doubling of fatigue life due to root radiusing. Research has additionally demonstrated that this potential applies to all fastener strength levels.

Roots of internal threads are not normally rounded. Mandatory root radiusing would mean use of taps with rounded crests and this would add unnecessarily to their cost. In a properly designed bolt/nut combination the nut is the stronger member of the team with the intent that should failure occur it will invariably be the externally threaded member.

4

Comparison of Thread Series

Thread series are groups of diameter/pitch combinations differing from each other by the number of threads per inch (or thread pitch) applied to a series of diameters.

The current Unified coarse thread series (UNC) is patterned on the thread series introduced by Whitworth in the mid-19th century. The relatively coarse pitches he selected were probably chosen as much for being those which lent themselves to the manufacturing skills of the day as for any other single reason. Over the years, as production capabilities improved, it became economically feasible to produce threads to greater degrees of accuracy and with finer pitches. Many special purpose thread series were developed and the one now known as the Unified fine thread series (UNF) enjoys considerable popularity. As screw thread technology advanced it became evident that just two thread series, coarse and fine, would be inadequate to efficiently satisfy all engineering applications. Consequently, a number of constant pitch thread series were added to the system. Such series have a single thread pitch which is common for all diameters in the series.

Metric thread systems evolved in a similar pattern and the current ISO system features a coarse and fine thread series supported by a number of constant pitch series.

Today there are 11 standard Unified thread series for inch products and 13 for metric. Fortunately, for fasteners only four have genuine importance—Unified coarse (UNC), fine (UNF) and 8-thread (8UN) in inch and metric coarse (M) in metric.

Few aspects of fastener engineering have been debated more intensely, and frequently with vehemence, than the pro and con merits of coarse versus fine threads. For a given diameter, fine threads are stronger in tension because of their larger tensile stress area. For the same conditions coarse threads have greater stripping strength over a given length of thread engagement. Because of their larger minor diameter fine threads develop higher torsional and transverse shear strengths. Stress concentration factor at thread roots decreases as the thread pitch increases. This suggests that, all else being equal, coarse threads should exhibit a superior fatigue resistance.

Fine threads are better for tapping thin walled members while coarse threads protect against deleterious loss of thread overlap due to nut dilation. Also fine threads are easier to tap in hard materials since less metal is being removed. On the other hand coarse threads are better for tapping brittle materials which tend to crumble or spall. Strength loss due to corrosion attack is not as harmful to coarse threads. Additionally coarse threads have slightly larger allowances in clearance fits (2A/2B) which means that thicker coatings and platings can be accomodated before thread adjustments have to be made.

Coarse threads can tolerate more abuse during handling and can be easier and quicker to assemble and disassemble. Fine threads provide better adjustment accuracy because of their smaller helix angles and require less torque to develop a given tension. However the actual difference is modest and usually can be ignored. Theoretically, fine threads have greater resistance to self-loosening under vibration. However, the tendency is so small that no credence can be given to fine versus coarse as a mechanism for maintaining bolt/nut tightness integrity.

Over the years the general debate has continued with no overwhelming support being generated for either series providing a reasonable indication that the technical merits and deficiencies are equally shared. The aerospace and automotive industries were originally strong proponents of fine threads. However in recent years there has been a visible shift toward increased usage of coarse thread fasteners. The motivation here is primarily to enjoy the economics of simplification rather than to advance any technical justification.

In the United States, fine thread fasteners smaller than No. 10 (0.190 in.) and larger than 1 in. are practically non-existent. Also for the sizes larger than 1 in., the 8-thread series (8UN) commands an equal popularity with coarse threads.

The ISO metric coarse thread series is uniquely positioned, in terms of its thread pitches, between the Unified coarse and fine inch threads. In other words, for comparable diameters, metric coarse thread pitches are finer than Unified coarse but coarser than Unified fine. And this makes the metric coarse series an ideal compromise with arguable technical advantages over either of the inch thread series. ISO metric fine threads have pitches considerably finer than the Unifed fine series and use of this thread series for metric fasteners is not recommended.

5

Classes and Grades of Thread Fit

Fit is a measure of looseness or tightness between mating threads. Classes of fit are specific combinations of tolerances and allowances as applied to external and internal threads.

For Unified inch series there are three classes for external threads (1A, 2A, and 3A) and three for internal threads (1B, 2B, and 3B). All are clearance fits, which means that these threads assemble without interference. Additionally, there is an interference fit, class 5, providing the tolerance for the external and internal threads which can assure positive interference. The higher the class number the tighter is the fit. The designator 'A' denotes an external thread while 'B' refers to an internal thread. Using these definitions class 1A/1B represents the loosest fit covered in screw thread standards while class 3A/3B is the tightest of the clearance fits.

For inch series threads, classes 1A and 1B are very loosely toleranced with an allowance applied to the external thread. These classes are suited where quick and easy assembly is a prime consideration. Class 1A/1B is standard only for coarse and fine threads for sizes of 1/4 in. and larger. These classes are rarely specified and in fact it is doubtful if more than one-tenth of 1% of all the fasteners produced in the United States have this class of fit specified for their threads.

Classes 2A and 2B are by far the most popular for inch series fasteners. Class 2A for external threads has an allowance while class 2B for internal threads does not. These classes however offer the optimum value of fit when balancing manufacturing convenience and economy, against the fastener performance.

Classes 3A and 3B are known to be best suited for precision fasteners such as socket cap and set screws, aerospace bolts and nuts, connecting rod bolts, and other high strength fasteners designed for use in safety-critical applications. However these classes have no specific allowance and are manufactured to restrictive tolerances.

About 90% of inch series fasteners have class 2A/2B threads. The remainder, except for a very small number, have 3A/3B classification which is the standard for all aerospace fasteners.

One of the unfortunate misconceptions about the mating threads is the belief that the tighter their tolerances and the closer the fit, the higher the quality of the assembly and the better its service performance. But, like an optical illusion, what appears to be an obvious truth is frequently false and designers giving selection priority to closer thread fits may unsuspectingly create assembly problems and cause unnecessary expense.

The differences between the classes of fit can be illustrated in Fig. 7. This example shows the tolerance principle for 1/2-13 UNC thread. As shown in Fig. 7 we have here three combinations of classes-of-fit based on pitch diameter with full thread profile for external and internal patterns. While the example is limited in Fig. 7 to 1/2-13 UNC thread similar fit characteristics apply to other Unified threads. Class 2A has a finite allowance, class 3A on the other hand does not. Class 3A tolerances are 75% those of 2A. Neither 2B or 3B classes for internal threads have an allowance while 3B tolerances are 75% those of 2B. Since class 3A does not have an allowance this means that in a true maximum material condition of both nut and fastener, there should be a complete contact of mating threads. In real life this kind of a combination is highly unlikely.

The corresponding tolerances for nut threads are 30% greater than those for fastener threads in all classes of fit. The tolerances are a function of pitch, fastener diameter, and length of engagement for Unified threads with some overlap between the three combinations of classes. This should be taken into account when planning the production. It is also of interest to note that regardless of thread class, working tolerances are extremely small. For instance, 2A and 3A classes for a 1/2-inch fastener are 0.0065 and 0.0048 inches, respectively. Values of tolerances for a variety of fast-

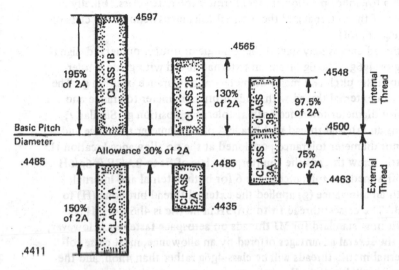

Figure 7 Example of tolerances for classes of fit on 1/2-13 UNC thread,

eners are given in standard handbooks and design manuals. It should also be
pointed out that the so-called maximum material condition defines the largest
bolt or smallest tapped hole in the nut that can be manufactured without exceeding permissible tolerances. Similarly, the minimum material condition
signifies the smallest permissible bolt thread diameter and the largest tapped
hole in the nut.

For ISO metric threads there is a comprehensive system of tolerance
grades and positions. The tolerance grade is the magnitude of the tolerance
and the tolerance position is the placement of the tolerance zone with respect
to the basic thread form. The ISO tolerance grade is designated by numerals
3 through 9 for external threads and 4 through 8 for internal threads, with
ascending numbers indicating increasing tolerances. Tolerance positions are
referred to by capital letters when applied to internal threads and lower case
symbols when dealing with external threads. The distance of the nearest end
of the tolerance zone from the basic size is called the funamental deviation
which is the ISO term for allowance. The tolerance position designated H,h
signifies zero allowance, G,g signifies a finite allowance with all other letters
F,f and lower being the allowances of increasing magnitude as the letters approach the start of the alphabet. The combination of a tolerance grade (nu-

meral) with a tolerance position (letter) forms a tolerance class. Finally a combination of the external and the internal tolerances denotes the class of thread fit, e.g., 6g/6H.

Metric thread classes may have either a single or double numeral designator, e.g., 6g or 4g6g. The 6g means an external thread with grade 6 tolerances for both the pitch and major diameters at a tolerance position g. The 4g6g means an external thread with grade 4 pitch diameter tolerance and grade 6 major diameter tolerance both at tolerance position g. Similarly, 4H5H means an internal thread with grade 4 pitch diameter tolerance and grade 5 minor diameter tolerance positioned at the no allowance location H.

For metric screw threads the counterpart class of fit to 2A/2B is 6g/6H which signifies medium tolerance grade 6 for both external and internal threads with an allowance (g) applied the external thread but none (H) to the internal. The closest thread fit to 3A/3B in metric is 4h6h/4H5H and this is the fit now standard for MJ threads on aerospace fasteners. However, because of the several advantages offered by an allowance, most close tolerance external metric threads will be class 4g6g rather than 4h6h, and the internal thread will be class 6H.

6

Effect of Thread Fit on Performance

In general different classes of thread fit may have different effects on the performance of a threaded fastener in a given mechanical joint. As mentioned previously, for instance, class 5 is a special thread indicating an overlap of material which must cause a finite amount of thread deformation before assembly can be accomplished.

Strengths of mating threads also depend on having adequate depth and length of thread engagement—depth being the amount of overlap in a transverse direction and length being the number of threads engaged longitudinally. Viewing the different classes of thread fit, it would appear that the load supporting capability of the closer toleranced tighter fit threads would be the stronger. But this is not necessarily true. Actually no tensile strength differential can be assumed between the loosest and tightest fits. The reason is that a large portion of the permitted tolerance zones for each class are common and there is no assurance that the threads of one class as actually manufactured contain more material than those of another. In the early 1940s the tensile strength experiments conducted at the Massachusetts Institute of Technology on threaded fasteners of different sizes and materials indicated that class 1 and class 3 fits were equally good and there was no reason to make fasteners with the tolerances closer than those of class 1. Later studies did not seriously challenge the validity of this observation which was based on the premise that the additional clearance of the looser fit might have indeed caused more uniform load distribution and lower contact stresses.

While different classes of fit may not exhibit any accountable differences in their tensile strengths, there is an appreciable difference in their ability to resist thread stripping. When an external thread strips it generally strips at the cylindrical plane generated by the minor diameter of the internal thread. Similarly, the failure location for an internal thread which strips is at the major diameter of the external thread. Thread shear areas are computed using minimum material conditions. Consequently, the no-allowance conditions and closer tolerances of class 3A/3B threads indicate more material to resist stripping in both the external and internal threads than is actually present in class 2A/2B. The difference can be quite significant. For example, 1/2-13 UNC 2A has a thread shear area of 0.779 sq. in. per in. of engaged length while 1/2-13 UNC 3A has 0.854 sq. in. per in., amounting to an increase of about 10%. For the internal threads, 1/2-13 UNC 2B has an area of 1.124 sq. in. per in. while 1/2-13 UNC 3B has 1.164 sq. in. per in., suggesting a 3.5% increase. This pattern is typical through the full range of fastener sizes for both coarse and fine threads.

For metric threads the difference is not as dramatic although there is a slight increase in the shear area of class 4g6g threads over those of class 6g. For example, M12 X 1.75 6g has a shear area of 19.0 sq. mm per mm of engaged length while for class 4g6g it is 19.8 sq. mm per mm. However, because both are usually mated with class 6H internal threads it is rather impractical to assume any stripping strength differential for these classes of metric threads. However in the case of inch series fasteners there may be an isolated application when the stripping strength differential between class 3A/3B and class 2A/2B threads could be important.

A large percentage of fasteners are plated or otherwise coated to give them corrosion protection or to enhance their aesthetic appearance. Platings and coatings have a finite thickness and add to the size of the part. Because of thread geometry the plating or coating adds four times its thickness to the thread diameters. Thread standards permit the use of the allowance for classes 2A, 6g, and 4g6g to accommodate plating thickness. For all but a few of the smaller diameter/pitch combinations the allowance is sufficient to accept commercial thicknesses of plating without any need for special processing. For thicker coatings however it may be necessary to undercut the external thread or overtap the internal thread to provide a free fit after plating. Because class 3A threads have no allowance for this purpose there is a distinct possibility that the threads after plating may not assemble. Consequently early attention must be given during the thread manufacture to assure a proper fit of the plated parts.

For fasteners used in elevated temperature service, generally above 500°F, it is desirable to provide a positive allowance between the mating threads in order to minimize galling and seizure as well as to furnish a space for lubrication. This is particularly important when the application requires occasional disassembly and reassembly. High cycle wrenching such as occurs on vehicle assembly lines, creates heat during fastener assembly because of the friction between the mating threads. The closer the fit, the greater the frictional effect.

Most fasteners experience considerable abuse and rough handling prior to use. External threads may be nicked or otherwise surface-damaged. While their strength properties may remain unaffected their ability to freely assemble may be impaired. Also the threads with an allowance can absorb more punishment than those which have no allowance provision.

Fasteners made of low and medium strength materials characterized by good ductility are advantaged by having class 2A or 6g threads. The reason here is that the allowance, coupled with the more liberal tolerances, provides breathing room between the mating threads to accommodate local yielding, thread bending, and other forms of deformation. When stressed the threads adjust to each other and the load is distributed more evenly. Conversely, fasteners of very high strengths with accompanying low ductility should have close fitting threads. The explanation for this is that when the fastener accepts its service load, the ability to adjust and deform is lessened. Consequently, any need to deform should also be eliminated to the extent possible by fabricating the threads as close to perfect form as production skills permit. This is also the principal reason why aerospace fasteners have 3A/3B threads classification.

It might appear that closer fitting threads would survive vibration without self-loosening better than those with an allowance fit. All else being equal this is possibly true. However there are other reliable and less costly techniques to control self-loosening than sheer dependency on thread fit.

The last consideration of course is cost. It is sufficient to say, the closer the tolerancing the higher the cost.

7

Criteria of Strength and Fracture

The four critical areas which give the assembled threads their load carrying capability are: (1) the effective cross-sectional area through the thread, known as the tensile thread area, which resists bolt fracture in tension; (2) the shear area of the external thread which resists the stripping off of the bolt thread; (3) the shear area of the internal thread which resists the stripping of the thread out of the nut or tapped hole; and (4) the cross-sectional area through the thread which resists transverse shearing through the bolt. This area is also used when computing torsional stresses and the resistance of the externally threaded fastener to twist off during tightening.

During the remainder of this discussion of strength factors the term bolt will represent the externally threaded fastener and the term nut will designate the internally threaded fastener or tapped hole.

The tensile stress and the thread shear areas establish the ability of the bolt/nut combination to resist failure during assembly and to safely support all externally applied tensile loads during its service life. Cross-sectional shear areas have importance only when analyzing the transversely applied service loads. The possibility of twist-off is relatively less important and of concern mainly when using tapping screws and other small size screws in tapped holes.

When a bolt/nut assembly is overtightened or statically overloaded in tension, failure will usually occur in one of the following three ways:

1. the bolt will break with the fracture occurring, almost invariably, through the threaded section and within one thread from the face of the nut or tapped hole;
2. the bolt threads will strip off (infrequent failure mode) when the nut material is much stronger than that of the bolt and when the length of engaged threads is less than about 60% of the nominal bolt diameter; and
3. the nut threads will strip out when the bolt material is appreciably stronger than that of the nut and there is an inadequate length of thread engagement.

It is of paramount importance in fastener selection that the designer assures, to the absolute degree possible, that in the event of a fastener failure due to the overtightening or overloading in service the failure will be through bolt fracture and not thread stripping. This is particularly important because recent trends have been to tension the bolts to progressively higher levels, frequently and intentionally beyond their yield strengths, in order to optimize the use of strength properties of the fastener, to enhance the fatigue behaviour of the joint, and to reduce any tendency of the fastener to loosen under service loads.

If, during tightening, the bolt breaks it is sudden and visible. Replacement is easy and the operator is alerted that immediate corrective action is necessary. Thread stripping, unfortunately, is an insidious type of a failure. It starts at the first stressed thread and, gradually, the remaining threads peel-off through the entire length of the engagement. It is a progressive failure which often takes several hours before the nut completely disengages from the bolt. A thread stripping failure can be initiated unsuspectingly, while visually the assembly appears satisfactory and there is no warning that the tightening practice needs adjustment. The key to preventing thread stripping failures is to provide an adequate length of thread engagement. The seemingly obvious solution then is to select thicker nuts or to provide deeper tapped holes. But, there is a practical point beyond which lengthening the thread engagement adds nothing but cost. Thicker nuts require longer bolts and both create unnecessary expense.

When a bolt/nut combination is axially loaded the bolt is stressed in tension while the nut is in compression. With rigid materials and threads of perfect form the load would be distributed uniformly throughout the engaged length with each thread supporting one full and equal share. However, fasteners are made of materials having elastic properties and rare indeed would be the event when both the external and internal threads simultaneously have perfect form.

As the load is applied the bolt stretches, lengthening the effective lead of its thread while the nut compresses and tends to shorten the lead of its thread. These deformations, while discreet, must balance each other both locally and over the entire length of thread contact. In practice there is a disproportionate distribution of the total load. The first engaged thread assumes a higher than average load with the remaining threads carrying successively lower loads. The last or top thread then experiences the lowest contact load. Analytical studies have shown that the load on the first thread can be over twice the average for all threads with the load carried by the last thread being considerably less than one-half the average. This is the principal reason why bolt fractures occur at the first thread within the nut or a tapped hole. Also, studies suggested that for the same length of thread engagement, the finer the thread pitch the higher the average load on the first thread. This is one of the contributing factors why fine pitch threads have a higher susceptibility to thread stripping failures.

Increasing the length of thread engagement much beyond one times the nominal bolt diameter is self-defeating. The reason is that with this number of engaged threads the portion of the total load carried by the top threads is quite low and can only increase if the first threads so grossly deform that their excessive share of the load is relieved and transferred to successive threads. At this point a progressive failure is likely imminent.

Deviations in thread form, particularly thread lead, may influence load distribution throughout the engaged length. For example, a bolt thread manufactured with a minus lead or a nut thread with a plus lead will help equalize the load distribution throughout the engaged length. The opposite condition aggravates it. However, for the gain to be accountable, special thread manufacturing is necessary. Computer analyses have shown that lead deviations within the limits accepted by standard attribute gages (GO and LO thread ring gages and GO and HI thread plug gages) have negligible effect on the strength of threads.

As the axially applied load increases the bolt elongates, its threads bend, the nut compresses, its threads also deform, and the nut walls dilate because the thread angle wedges the contacting thread apart radially. This radial force is resisted by hoop stresses in the nut. The lower the strength of the nut material and the thinner the wall section, the greater the dilation.

Controlling nut dilation is important because dilation occurs at the nut bearing face which is the location of the most highly stressed thread. As the nut moves radially the depth of engagement of this thread is reduced which in turn decreases the thread stripping area of both the bolt and nut. This process increases unit shear stressed and is possibly the determining factor whether the bolt fractures or the threads strip. The finer the thread pitch the more aggravated this situation, which is another reason why fine thread (UNF) nuts experience more stripping failures than coarse (UNC). Hex nuts with widths across flats less than one and a half times their nominal thread diameter should be viewed with caution. In the case of tapped holes however the surrounding material is comparatively massive and the dilation as a strength reducing factor may be ignored.

When the relative strengths of the bolt and nut threads are approximately the same the axial load will cause both threads to bend intensifying nut dilation and the effective shear areas of both threads will decrease. When a thread stripping failure results it is difficult to determine whether the bolt thread or that of the nut failed. If the strengths of the bolt threads are significantly greater than those of the nut (the majority of all bolt/nut combinations) the bolt threads will not distort as readily and will constrain the nut threads from bending even though the nut material may have a much lower yield strength. If, under these conditions, a thread stripping failure occurs the nut will strip-out cleanly on the cylindrical plane generated by the major diameter of the bolt thread. Similarly, when the strength of the nut threads exceeds that of the bolt threads, thread bending and distortion become constrained, and if the assembly fails by thread stripping the bolt thread will strip-off at the cylindrical plane generated by the minor diameter of the nut thread. Generally, the closer the nut and bolt thread strengths are to each other, the lower the thread stripping failure load regardless of which one strips. If there is a strength disparity the failure load will be slightly higher. This is the reason why nuts tested using hardened mandrels exhibit higher failure loads than when tested using bolts of comparable strength properties.

An interesting phenomenom is the influence that the number of exposed bolt threads within the grip (the stressed length) has on the ultimate tensile strength of the fastener. If a bolt/nut combination is tested to failure in tension and the nut is positioned anywhere along the bolt thread length with at least four or more threads exposed between the nut bearing face and the bolt thread runout, the tensile strength of the bolt, i.e., the maximum load it can support prior to or at failure, will remain essentially unchanged. As the nut is brought closer to the thread runout, thus reducing the number of exposed threads, the tensile strength of the bolt increases and will be as much as 20% greater when the bolt is tested with the nut advanced as far as the thread run-out permits. While the bolt's apparent tensile strength is increasing, the stripping strength of the bolt and nut threads are decreasing because necking of the bolt is now occurring in the threaded length which is engaged by the nut. This reduces depth of thread engagement and drastically increases thread shear stresses. Conceivably, it is possible to change the failure mode from bolt fracture to thread stripping just by reducing the number of exposed threads within the grip.

In order to assemble and tighten the fasteners one or both mating components must be torqued which further complicates bolt/nut thread strength relationships. Torquing a bolt or nut induces a combined tensile and torsional stress in the bolt such that if tightening is continued to bolt failure the observed tensile strength would be only about 85% of the tensile strength of the same bolt if tested to failure in axial tension. The torsional stress results from the frictional resistance to rotation of the contacting threads. While overcoming frictional resistance to rotation, the threads are also free to move radially. And, if tightening is by the nut the frictional resistance between its surface and the material being joined is broken. Both conditions facilitate nut dilation.

Thus, when fasteners are tightened by torquing, their breaking strength and resistance to thread stripping are lowered. However, tests have proven that during assembly the stripping failure mode is marginally advantaged over bolt fracture, even though the fasteners may be liberally lubricated which reduce friction and adds to the possibility of nut dilation. As soon as tightening is discontinued the torsional stress component in the bolt immediately dissipates and its tensile strength is again 100% of its capacity when tested to failure in tension. Unfortunately, nut dilation induced by tightening does not bounce back and remains throughout the service life.

It is theoretically possible to design a threaded joint with a 100% assurance of bolt fracture as the failure mode. However, the economics are severely punishing because of the material waste. ISO nuts were designed (widths across flats, thicknesses, countersinks, proof loads, and hardnesses) using a simple but highly practical criterion. It was assumed that when in a mass production the nuts were sufficiently overtightened to the point of failure, at least 10% of these failures were by bolt fracture. This constituted an adequate red-flag warning to the operator that immediate corrective action was necessary. Once installed then the failure mode of a properly selected bolt/nut combination, if overloaded in tension, should always be by bolt fracture.

Table 1   Thread Stress Areas for Inch Series Fasteners

Nominal size, Threads per Inch and Thread Series	Tensile Stress Area (sq in.) (AS)	Thread Root Area (sq in.) (AR)	Thread Stripping Areas (sq in. per in. of engagement)			
			External Thread (AS_s)		Internal Thread (AS_n)	
			Class 2A	Class 3A	Class 2B	Class 3B
No. 0-80 UNF	0.00180	0.00151	0.0673	0.0748	0.106	0.116
1-64 UNC	0.00263	0.00218	0.0835	0.0914	0.133	0.144
2-56 UNC	0.00370	0.00310	0.101	0.109	0.162	0.174
3-48 UNC	0.00487	0.00406	0.118	0.128	0.191	0.204
4-40 UNC	0.00604	0.00496	0.138	0.147	0.221	0.235
5-40 UNC	0.00796	0.00672	0.161	0.172	0.248	0.263
6-32 UNC	0.00909	0.00745	0.180	0.190	0.281	0.296
8-32 UNC	0.0140	0.0120	0.226	0.238	0.334	0.354
10-24 UNC	0.0175	0.0145	0.263	0.275	0.401	0.420
10-32 UNF	0.0200	0.0175	0.275	0.289	0.389	0.405
12-24 UNC	0.0242	0.0206	0.312	0.327	0.458	0.479
1/4-20 UNC	0.0318	0.0269	0.368	0.385	0.539	0.563
1/4-28 UNF	0.0364	0.0326	0.373	0.403	0.521	0.549
5/16-18 UNC	0.0524	0.0454	0.470	0.502	0.682	0.710
5/16-24 UNF	0.0580	0.0524	0.479	0.520	0.663	0.696
3/8-16 UNC	0.0775	0.0678	0.576	0.619	0.828	0.860
3/8-24 UNF	0.0878	0.0809	0.578	0.644	0.800	0.837
7/16-14 UNC	0.106	0.0933	0.677	0.734	0.981	1.01
7/16-20 UNF	0.119	0.109	0.685	0.761	0.908	0.991
1/2-13 UNC	0.142	0.126	0.779	0.854	1.12	1.16
1/2-20 UNF	0.160	0.149	0.799	0.887	1.08	1.13
9/16-12 UNC	0.182	0.162	0.893	0.974	1.27	1.32
9/16-18 UNF	0.203	0.189	0.901	1.02	1.23	1.29
5/8-11 UNC	0.226	0.202	0.998	1.09	1.42	1.47
5/8-18 UNF	0.256	0.240	0.998	1.13	1.37	1.43
3/4-10 UNC	0.334	0.302	1.21	1.34	1.72	1.78
3/4-16 UNF	0.373	0.351	1.23	1.38	1.66	1.73
7/8- 9 UNC	0.462	0.419	1.43	1.58	2.03	2.09
7/8-14 UNF	0.509	0.480	1.44	1.63	1.96	2.03
1- 8 UNC	0.606	0.551	1.66	1.82	2.33	2.40
1-12 UNF	0.663	0.625	1.66	1.87	2.27	2.35
1-14 UNS	0.680	0.646	1.67	1.89	2.23	2.33

Table 1   (Continued)

Nominal size, Threads per Inch and Thread Series	Tensile Stress Area (sq in.) (AS)	Thread Root Area (sq in.) (AR)	Thread Stripping Areas (sq in. per in. of engagement)			
			External Thread (AS$_s$)		Internal Thread (AS$_n$)	
			Class 2A	Class 3A	Class 2B	Class 3B
1-1/8- 7 UNC	0.763	0.693	1.88	2.04	2.65	2.72
1-1/8- 8 UN	0.790	0.728	1.89	2.07	2.63	2.70
1-1/4- 7 UNC	0.969	0.890	2.11	2.30	2.94	3.02
1-1/4- 8 UN	1.000	0.929	2.12	2.33	2.92	3.00
1-3/8- 6 UNC	1.16	1.05	2.34	2.52	3.27	3.35
1-3/8- 8 UN	1.23	1.16	2.34	2.58	3.21	3.30
1-1/2- 6 UNC	1.41	1.29	2.58	2.77	3.57	3.65
1-1/2- 8 UN	1.49	1.41	2.57	2.84	3.50	3.61
1-5/8- 8 UN	1.78	1.68	2.80	3.10	3.79	3.91
1-3/4- 5 UNC	1.90	1.74	3.04	3.24	4.20	4.30
1-3/4- 8 UN	2.08	1.98	3.03	3.35	4.08	4.21
1-7/8- 8 UN	2.41	2.30	3.25	3.63	4.37	4.50
2-4-1/2 UNC	2.50	2.30	3.53	3.72	4.83	4.93
2-8 UN	2.77	2.65	3.48	3.86	4.66	4.81
2-1/4-4-1/2 UNC	3.25	3.02	4.02	4.23	5.44	5.55
2-1/4-8 UN	3.56	3.42	3.93	4.37	5.24	5.40
2-1/2-4 UNC	4.00	3.72	4.50	4.70	6.07	6.20
2-1/2-8 UN	4.44	4.29	4.38	4.87	5.81	6.00
2-3/4-4 UNC	4.93	4.62	4.99	5.22	6.68	6.82
2-3/4-8 UN	5.43	5.26	4.83	5.38	6.39	6.60
3-4 UNC	5.97	5.62	5.48	5.74	7.29	7.44
3-8 UN	6.51	6.32	5.28	5.89	6.95	7.20
3-1/4-4 UNC	7.10	6.72	5.97	6.26	7.90	8.06
3-1/4-8 UN	7.69	7.49	5.73	6.40	7.53	7.79
3-1/2-4 UNC	8.33	7.92	6.47	6.77	8.51	8.68
3-1/2-8 UN	8.96	8.75	6.18	6.90	8.10	8.39
3-3/4-4 UNC	9.66	9.21	6.95	7.29	9.11	9.31
3-3/4-8 UN	10.34	10.11	6.61	7.41	8.67	8.98
4-4 UNC	11.08	10.61	7.44	7.81	9.71	9.92
4-8 UN	11.81	11.57	7.07	7.91	9.24	9.57

Table 2   Thread Stress Areas for Metric Series Fasteners

Nominal Thread Diameter and Thread Pitch	Tensile Stress Area (sq mm) (AS)	Thread Root Area (sq mm) (AR)	Thread Stripping Areas (sq mm per mm of engagement)	
			External Thread - 6g (AS$_g$)	Internal Thread - 6H (AS$_n$)
M1.6 X 0.35	1.27	1.03	1.87	2.64
M2 X 0.4	2.07	1.72	2.45	3.44
M2.5 X 0.45	3.39	2.88	3.18	4.48
M3 X 0.5	5.03	4.34	3.90	5.55
M3.5 X 0.6	6.78	5.81	4.67	6.61
M4 X 0.7	8.78	7.50	5.47	7.77
M5 X 0.8	14.2	12.3	7.07	10.0
M6 X 1	20.1	17.3	8.65	12.2
M8 X 1.25	36.6	31.9	12.2	16.9
M10 X 1.5	58.0	50.9	15.6	21.5
M12 X 1.75	84.3	74.3	19.0	26.1
M14 X 2	115	102	22.4	31.0
M16 X 2	157	141	26.1	35.6
M20 X 2.5	245	220	33.3	45.4
M22 X 2.5	303	276	37.0	50.0
M24 X 3	353	317	40.5	55.0
M27 X 3	459	419	46.2	62.0
M30 X 3.5	561	509	51.6	70.1
M36 X 4	817	745	63.1	84.1
M42 X 4.5	1120	1030	74.3	99.2
M48 X 5	1470	1350	85.8	114
M56 X 5.5	2030	1870	101	134
M64 X 6	2680	2480	117	154
M72 X 6	3460	3240	133	173
M80 X 6	4340	4090	149	193
M90 X 6	5590	5310	169	217
M100 X 6	6990	6680	189	241

Unfortunately, as indicated previously the same situation is not true for all inch fasteners and in some combinations it is entirely possible that the nut thread will strip before the bolt breaks both during the assembly and in service. Tables 1 and 2 list the thread stress areas for inch and metric series fasteners. The formulas for computing these values are as follows:

$$\text{Tensile stress area, inch series} = AS = 0.7854 \left(D - \frac{0.9743}{n} \right)^2$$

where

AS = stress area, sq. in.

D = nominal size (basic major diameter of thread), in.

n = number of threads per inch

$$\text{Tensile stress area, metric series} = AS = 0.7854(D - 0.9382p)^2$$

where

AS = stress area, sq. mm

D = nominal size (basic major diameter of thread), mm

p = thread pitch, mm

$$\text{Thread root area, inch series} = AR = 0.7854 \left(D - \frac{1.3}{n} \right)^2$$

where

AR = root area, sq. in.

D = nominal size (basic major diameter of thread), in.

n = number of threads per inch

$$\text{Thread root area, metric series} = AR = 0.7854(D - 1.3p)^2$$

where

AR = root area, sq. mm

D = nominal size (basic major diameter of thread), mm

p = thread pitch, mm

Thread stripping area of external thread =

$$AS_s = \frac{3.1416 \, Le \, Kn \, max}{p} \, [0.5p + 0.57735(Es \, min - Kn \, max)]$$

where

AS_s = thread stripping area, sq. in. or sq. mm

Le = length of engaged threads, in. or mm

Kn max = maximum minor diameter of internal thread, in. or mm

p = thread pitch, in. or mm = $1/n$ where n equals threads per inch

Es min = minimum pitch diameter of external thread, in. or mm

Thread stripping area of internal thread =

$$AS_n = \frac{3.1416 \, Le \, Ds \, min}{p} \, [0.5p + 0.57735(Ds \, min - En \, max)]$$

where

AS_n = thread stripping area, sq. in. or sq. mm

Le = length of engaged threads, in. or mm

Ds min = minimum major diameter of external thread, in. or mm

p = thread pitch, in. or mm = $1/n$ where n equals threads per inch

En max = maximum pitch diameter of internal thread, in. or mm

8

Selection of Materials

Fasteners are made of dozens of different materials such as ferrous, non-ferrous, and non-metallic (plastics). And necessarily so, because of the almost limitless combinations of service conditions in which fasteners are expected to perform.

Selection of exactly the right material for a fastener in a specific application may well appear to be awesome. However, in practice, any engineer with a fundamental knowledge of the properties, capabilities, and the limitations of various materials should not find it too difficult to zero-in on the right choice of base material. For example, if the fastener's main function is to deliver strength, think of steel. If the atmosphere is corrosive, consider steel with a protective coating. For a particularly severe corrosive environment select stainless steel or one of the non-ferrous alloys. If magnetic permeability is important, by-pass conventional steel and choose austenitic stainless steel, aluminum, copper, or plastic alloy. Where high electrical conductivity is a major requirement, turn to aluminum or copper. When resistivity is the prime consideration give thought to a non-metallic material. For a weight saving task select aluminum. When, however the weight-saving feature is coupled with a need for high strength the clear choice is titanium. And for the case of high and low temperature service investigate stainless steels and the super alloys.

The main point here is that the engineer does not need to be an experienced metallurgist. His responsibility is to evaluate the intended service and to quantify the design parameters the fastener must satisfy. With this information at hand the selection of the proper material should not be a problem.

Within each base material there are usually a number of analyses of slightly varying chemical composition that could be picked as the specific raw material for the fastener. By modestly juggling manufacturing and processing techniques it is quite feasible to use any one of them and produce fasteners which totally satisfy the needed performance properties. The design engineer is not expected to be familiar with all the subtle material, manufacturing, and processing options. The detailed options are reserved for the fastener producer who is in the ideal position to make the final choice of the raw material technology.

For most fastener applications the engineer defines the performance and specifies the base material and environment leaving the final choice of material analysis to the manufacturer's judgment within the constraint of schedules and economy. The more latitude permitted the producer, the more efficiently he can employ his equipment and processing skills. In general, fastener cost relates directly to the broadness of choice permitted the manufacturer. Over-specifying stifles ingenuity and creates cost penalties.

Perhaps a logical continuation, before getting into a fuller discussion of the properties of fasteners made from the different base materials, might be to review some of the more important design considerations influencing fastener material selection.

The cost of the raw material in fasteners represents from 30 to 80% of their manufactured cost. By far the least expensive material is low carbon steel. As carbon content increases and as alloying elements are added the cost of steel increases but medium carbon and low alloy steels are still by several times the lowest cost of all the metallic materials. For example, and in the most general of terms, assuming the cost of medium carbon steel as unity, stainless steels (300 series) cost two and a half times as much, copper alloys five times, aluminum alloys six times, A-286 12 times, Monel 20 times, titanium 75 times, and the super-heat resistant and low temperature alloys up to 100 times. The design engineer must be alert to raw material costs and select a material no more expensive than that which provides the necessary combination of performance requirements.

9

Practical Strength Limits

It would indeed be a unique service application in which the fastener did not support or transmit some form of externally applied load. And when fastener strength is the single design consideration, there is no need to look beyond steel. As indicated previously, this material is by far the most economical and versatile. In fact raw material costs are such that choices other than steel should only be considered when some special requirement in combination with strength, dictates that selection. Detailed evaluation of carbon steel fasteners is given in Chapter. 15.

Low-carbon steel fasteners have tensile strengths of 60,000 psi; medium carbon heat-treated fasteners 120,000 psi; low alloy steel 150,000 psi and higher; and some of the super alloys have strengths of 260,000 psi and even higher. In fact, fasteners with strengths exceeding 400,000 psi are technologically possible, although designing with and using such fasteners is no game for the amateur. In most engineering applications other than aerospace, there is seldom any need to consider fasteners with tensile strengths exceeding 180,000 psi.

Each of the three basic types of *stainless steel*—austenitic, ferritic, and martensitic—have distinctly different strength properties. Further informa- on the use of various stainless steels is included in Chapters 22 through 25.

Austenitic grades, such as 303, 304, 305, 316, and 321, are non-heat treatable; however their strength properties can be improved through cold working and strain hardening techniques. Generally, solution annealed fasteners of these grades have tensile strengths of 75,000 psi and cold worked fasteners up to 90,000 psi. When strain hardened however, depending on fastener size, strengths of 125,000 psi are possible. The ferritic grades, such as 430 and 430F, do not respond to heat treatment and their tensile strength is limited to about 70,000 psi. On the other hand fasteners of martensitic stainless steels, such as 410, 416 and 431, are heat treatable and can have strengths as high as 180,000 psi. Specific applications of austenitic and martensitic stainless steels to fastener production are discussed in Chapters 23 and 25.

On a quantity basis there are more *aluminum and aluminum alloys* fasteners produced than of any other non-ferrous material. The reasons include aluminum's low weight, its high strength-to-weight ratio; good corrosion resistance in most environments, excellent thermal and electrical conductivity, and an adequate performance in low temperatures. It also has respectable strength properties. There are literally dozens of aluminum alloys and tempers, and fasteners can be produced from most of them. Tensile strengths may range from pure aluminum with its 13,000 psi to the most popularly used alloys for fasteners, 2024 and 7075, with strengths slightly higher than 60,000 psi. Special features of aluminum alloys are summarized in Chapter 27.

None of the *copper base alloys* (brass and bronze) respond to heat treatment and any strength appreciation of the fastener over that of its raw material can only be achieved through cold working. Since, however, many copper alloys must be stress relieved after forming in order to eliminate embrittlement, the resulting fastener strengths usully parallel those of the base alloy. The tensile strength range for the commonly used copper alloys spans from 50,000 psi for No. 462 naval brass to 105,000 psi for No. 630 aluminum bronze. Applications of copper base alloys are evaluated in Chapter 26.

Nickel base alloys have excellent strength properties which in combination with their superior corrosion resistance, toughness, and unparalleled performance in high and low temperature extremes, gives these materials a popularity limited only by their relative high cost. For commercial fasteners the most frequently specified alloys are of the nickel-copper type (Monel is an example) with a tensile strength of 80,000 psi and heat treatable nickel-copper-aluminum (K-Monel for example) with a tensile strength close to 130,000 psi. Further discussion of nickel based alloys is given in Chapter 28.

Titanium, because of its superior strength-to-weight ratio, enjoys universal acceptance by the aerospace, missile, and some chemical processing industries, where a unique combination of mechanical properties justifies its high cost. Bolts of the two workhorse alloys, Ti-4Mn-4Al and Ti-6Al-4V, have tensile strengths of 150,000 psi and fasteners with strengths up to and even slightly higher than 200,000 psi are now available. Additional information on titanium alloys is included in Chapter 29.

There are a number of valid reasons why a fastener of a *nonmetallic* material could be the ideal choice for an engineering application if it were not for a limited strength. Compared to metallic fasteners, plastic fasteners are not strong although, surprisingly, the strength-to-weight ratio of some plastics relates favorably to that of a low carbon steel. About the best that can be expected from plastics, including the widely used nylon, is a tensile strength on the order of 10,000 psi. Further discussion of the application of nonmetallics to fastener field is given in Chapter 31.

10

Materials for Corrosive Environment

Corrosion is the deterioration of metal in its environment. It is insidious, wasteful, and unsightly. It is difficult and costly to control, and frequently is the cause of fastener failure. Corrosion costs American industry and consumers hundreds of millions of dollars every year, with absolutely no benefit or return.

When analyzing a fastener application one of the first questions the designer should ask is what is the service environment and is there a possibility the fastener will be subject to corrosive attack? Seldom will the answer be no. Any yes answer obligates the engineer to identify the character of the exposure, anticipate the possible consequences, and determine the most effective counter-measures. Because of the extreme importance of corrosion, not only atmospheric exposures but galvanic actions, high temperature oxidation, and dangerous stress corrosion embrittlement should be considered.

Steel fasteners with some type of protective plating or coating perform well in most atmospheric environments. Generally, the thicker the plating or coating, the more effective is the protection against corrosive attack. But, for thicker platings, such as galvanizing, some modification in the thread limits of mating threaded parts must be made to assure ease of assembly. Also, as strength of externally threaded parts increases, their susceptibility

to stress corrosion embrittlement increases and steel fasteners with hardnesses of Rockwell C40 and higher present the engineer with a genuine
problem.

Several years ago a new family of steels, having excellent atmospheric
corrosion resistance, was introduced. These materials are now quite popular in the construction of atmospherically exposed steel structures such as
bridges, buildings, and transmission towers. Commonly known as weathering steels, these low alloy steels have a significant content of copper which
helps build a protective oxide surface film on the bare steel when it is exposed to the atmosphere. After a few years the film stabilizes and effec-
tively combats further corrosive attack.

In more aggressive environments, e.g. marine, chemical plants, paper
mills, or refineries regular steel fasteners, even though protected, may not
be entirely suitable and thought must be given to stainless steels and non-
ferrous alloys. There are many to choose from and each has its own corrosion fighting identity. In most applications any one of several materials
could provide sufficient corrosion resistance although on rare occasions the
choice may narrow to a single alloy suitable for a specific environment.
For example, Type 316 stainless steel performs well in contact with chro-
mic acid, other stainless steels, copper, aluminum, and nickel alloys do not.

Corrosive attack may come from many directions. There are usually
several options of the most effective means to combat this environment.
Corrosion of any type is not a design consideration to be treated casually.
The designer is well advised to give the problem his best thinking and, if in
doubt as to the appropriate action, is urged to consult an experienced metallurgist.

11

Materials for High Temperatures

Most properties of materials are temperature sensitive. The principal effect of increasing temperature on metallic fasteners is a decline in their strength properties. When the temperature becomes high enough, about 500°F and up, other problems begin to occur, such as breakdown of coatings, high temperature oxidation and/or corrosion, possible incompatibilities of coefficients of expansion, and galling and seizing.

Few of the non-ferrous metals are suitable for high temperature service and non-metallics should generally be avoided. Consequently, first consideration should be given to ferrous materials, ranging from the medium carbon steels through the super alloys. For service temperatures up to 450°F (230°C) the commercial grades of medium carbon and alloy steels, frequently used in fastener manufacture, will perform satisfactorily. From 450°F to about 900°F (480°C) stainless steels and the chrome-moly steels in the AISI 4100, 8600 and 8700 series (covered by ASTM A193) are adequate. For higher temperature service one of the specially developed heat resistant super alloys must be used. For instance A-286 and Inco 718 exhibit reasonably good strength properties to about 1200°F (650°C) while the nickel-cobalt alloys such as M252, Waspalloy, Udimet, and Rene 41 are good up to 1600°F (870°C). Beyond this temperature the designer must turn to refractory metals with their high alloy content of columbium, molybdenum, tantalum, or tungsten. Some of the tungsten alloys have performance capability in temperatures exceeding 3000°F (1650°C).

Perhaps the most severe problem associated with elevated temperatures, in addition to fastener strength loss, is the instability of platings and coatings. For example, conventional cadmium platings are dependable only to about 450°F after which they begin to melt. For higher temperatures, platings with a high nickel content are reasonably effective in providing corrosion protection and the important reduction of galling and seizing. Above 900°F to about 1400°F silver plating helps. However for higher temperatures it is difficult to find any plating that will survive and function for any appreciable length of time.

The number of applications which experience sustained temperatures exceeding 900°F are relatively few. Here the engineer faces very special and complex problems justifying the guidance of a qualified and knowledgeable metallurgist. Additional information on high temperature alloys is given in Chapter 30.

12

Low Temperature Materials

In northern climates with occasional temperature drops to −30°F, most metallic fasteners will function satisfactorily. However, with prolonged exposures to subzero environments, such as those experienced in Arctic regions, the prudent designer should give careful attention to the selection of the fastener material.

Generally, as temperature drops, the tensile, yield, and shear strengths of most metallic materials increase. Unfortunately metal becomes brittle as its ductility and toughness decreases. Any material having notch sensitivity, such as low carbon steels, is particularly unsatisfactory for use in low temperature applications.

ASTM A320 defines a number of alloys which have low temperature service capability. AISI 4140 and 4037 may be suitable for atmospheres which are not overly severe with the nickel-rich AISI 4340 and 8740 series suitable for temperatures as low as −100°F. The well known austenitic stainless steels (18-8 series) will perform exceptionally well down to about −300°F. Exposure of these nickel chromium steels to extremely low temperatures has no significant effect on their strength or impact properties either for short or prolonged periods of time. Their use at lower temperatures however should be viewed with a degree of caution.

Other materials exhibiting satisfactory low temperature behavior include 2024-T4 aluminum, many of the titanium alloys, some of the copper and brasses, and even one or two of the non-metallics in the teflon category.

With the advent of the space age and the need to handle and contain liquid gas propellants, fasteners with a capability to perform in the cryogenic temperature range (−250°F down to −450°F) had to be developed. As can be imagined, developing and testing fasteners for such extremely low temperatures is difficult and enormously expensive. The performance at cryogenic temperatures is not entirely predictable and each candidate alloy must be researched individually. Among the proven alloys to date are Inco 718, Unitemp 212, A-286, Rene 41, and Waspalloy. It is interesting to note that these good performers are also among the best at extremely high temperatures. Additional material on low temperature materials is included in Chapter 30.

13

Mechanical Properties of Fasteners

Physical, mechanical, and performance characteristics are material-process related properties which collectively give the fastener its service capability. *Physical properties* are inherent in the raw material and remain unchanged, or with only slight alteration, in the fastener following its manufacture. Such physical properties as density, coefficient of thermal expansion, electrical resistance, thermal conductivity, and magnetic susceptibility all have importance and frequently dictate fastener material selection. *Mechanical properties* identify the reaction of a fastener to applied loads. Rarely are the mechanical properties of the fastener equal to those of the raw material from which it is made. Properties such as tensile strength, yield strength, hardness, and ductility are all subject to dramatic changes depending on the choice of manufacturing methods and metallurgical treatments. Selection of the raw material establishes a basic plateau of properties while the final combination of mechanical properties of the fastener itself are obtained during the manufacture and post-treatment processes. *Performance properties* are functional design features built into the fastener in order to satisfy criteria of the service application. Properties such as locking ability, prevailing torque, sealing, and driving torque are generally gained through the manufacturing control of dimensional features. Choice of the material and its metallurgical treatment influence the magnitude rather than the existence of a particular performance property.

Mechanical properties generally relate to strength characteristics. Fastener strength grades of most metallic materials are well standardized and are defined in a number of regulatory documents, such as those published by ASTM, SAE, ISO, and IFI. Test procedures for determining mechanical properties are equally well standardized. The best source for this information is ASTM F606.

Tensile strength is the maximum tension applied load a fastener can support prior to or coincident with its fracture. Tensile strengths (and this also applies to yield strengths and proof loads) are normally expressed as a stress-pounds per square inch (psi) for inch series fasteners and megapascals (MPa) for metric. To convert to a load (lbs) multiply the stress (psi) by the stress area (sq. in.) of the thread as given in Table 1. Similarly, for metric fasteners multiply the stress (MPa) by the thread stress area (mm^2) as given in Table 2 and divide by 1000 to derive the load in kilonewtons (kN). Some of the basic conversion units are given in Chapter 40.

It is most desirable to test fasteners full size. This is quite practical for externally threaded products having specified tensile strengths of 100,000 lbs (450 kN) or less, as most manufacturers have testing equipment of at least this capacity. For stronger fasteners it may be necessary to use test specimens of reduced size, machined from the actual fastener under test. The recommended dimensions of test pieces are given in ASTM F606. Because of necessary test fixturing and of the criticality of thread length exposure to the stressed region (as explained in Chapter 7 on fastener strength criteria), it is impractical to perform tensile tests on fasteners with lengths shorter than 2.25 times their nominal diameter. Short fasteners, therefore, are acceptance-tested on the basis of hardness (rather than strength).

Hex and hex flange screws and bolts, studs, and hex socket cap screws are wedge tensile tested. Other externally threaded fasteners are tested in an axial mode when determining their tensile strengths. The only difference is that in the wedge tensile test a hardened washer with a beveled surface is placed under the head of the test fastener. As the load is applied the wedging action induces a severe bending stress, concentrated at the junction of the head and the shank. To be acceptable the fastener must without failure, support, its specified minimum tensile strength. It is also important that the test fracture occurs at a point other than the junction of head and shank. Such a test demonstrates ductility and structural integrity at this critical location. The wedge angle varies between 4 and $10°$ depending on fastener size, strength grade, head style, and closeness of threading to the underside of the head.

Yield strength is the tension applied load at which the fastener experiences a specified amount of permanent deformation. In other words, the material has been stressed beyond its elastic limit and has entered its plastic zone. Yield strengths of machined test specimens are easily determined because of their uniform cross-sectional area throughout the stressed length. However, yield strength properties of test specimens do not always parallel those of the full size product from which they are taken because the beneficial effects of cold working may be completely lost when the test specimen is machined from the parent product. Unfortunately, it is extremely difficult to test full size fasteners for yield strength because of the different strain rates in the fully threaded portion, the thread runout, and the unthreaded shank which comprise the stressed length. For this reason the "proof load" concept was introduced as a convenient technique for testing a fastener's deformation characteristics.

Proof load is the tension applied load which the fastener must support without evidence of any permanent deformation. Proof load is an absolute value. It is neither a maximum or a minimum. For most fastener strength grades proof loads are established at approximately 90 to 93% of the expected minimum yield strength of the fastener material. In the testing procedure the overall length of the fastener is measured, the proof load applied, released, and the length remeasured. To be acceptable, the after-loading length must be the same as the original, within a small tolerance permitted for measuring error. The absence of permanent deformation proves that the yield point of the fastener has not been exceeded. Proof loads have added importance as they are frequently used as design values in joint analysis and fastener selection.

Hardness is a measure of a material's ability to resist abrasion and indentation. The tremendous importance of fastener hardness as a specification requirement is that hardness testing is quick, easy, and nondestructive. For steel fasteners there is also a close correlation between the hardness and tensile strength parameters. For nonferrous and stainless steel fasteners there is a similar pattern but the relationship is not as precise as that for steel. Hardness levels of fasteners are usually expressed in terms of Rockwell, Brinell, or Vickers (Diamond Pyramid) numbers. Rockwell has several scales (A, B, C, 15N, 30N) while Brinell and Vickers have one each. Hardness conversion tables with the corresponding tensile strength values for steel are covered in ASTM A370 and SAE J417.

In most fastener strength grade standards a hardness range is specified for the base metal. The minimum corresponds to the specified minimum tensile strength. The maximum represents a level of hardness beyond which the fastener might be unacceptably brittle. For routine hardness testing measurements are made on the top or side of the head, on the shank, or on the end of the product. For referee testing, hardness is measured at mid-radius of a cross-section through the threaded length at a distance of one diameter from the end of the product.

For quenched and tempered steel fasteners it is desirable to measure the hardness at the surface in the threaded section and compare this surface hardness with that of the base metal. A significantly lower surface hardness signals the presence of decarburization, i.e., a soft surface layer due to carbon loss during heat treatment. If the hardness is on the high side there is a distinct possibility that the surface is carburized, i.e., the skin on the fastener is harder, and consequently more brittle, than the underlying metal. Both conditions of course represent a potential risk of inferior performance by the fastener in its service application.

Ductility is the ability of a material to deform before it fractures. A rubber band is very ductile, glass is not. The common characteristics of ductility are determined by measuring the elongation and reduction of area of the machined test specimens under load. However, when testing a full size fastener these criteria are largely meaningless because yielding and eventual fracture normally occur in the threaded length. The process of measuring the magnitude of elongation and the reduction of area in a threaded section is totally impractical. The two most meaningful indicators of satisfactory ductility have been observed when the specified maximum hardness of the fastener has not been exceeded and when the wedge tensile test has shown the required tensile strength. A reasonable indication of a fastener's ductility is the ratio of its specified minimum yield strength to the relevant minimum tensile strength. The lower this ratio the more ductile the fastener will be. Conversely, the higher the ratio the lower is the fastener's ductility.

Toughness is a material's ability to accept punishment in the form of impact and shock loading. The measure of toughness is impact strength, which is a determination of energy absorbed by a standard test specimen when struck sharply under controlled test conditions. Impact strength testing of non-aerospace fasteners is rarely a specification requirement with the one notable exception of carbon and low alloy steel fasteners intended for low temperature service as covered in ASTM A320.

Shear strength is the maximum load, applied normal to a fastener's axis that the fastener can support prior to its failure. Single shear is load occurring in one transverse plane which tends to cut the fastener into two pieces. A double shear condition exists when the load is applied in two planes so that, at failure, the fastener could be cut into three pieces. For aerospace fasteners, shear strength and shear testing is a frequently specified requirement. For industrial fasteners, with the exception of transmission tower bolts and blind rivets, it is seldom a requirement. For blind rivets the shear testing procedure is well standardized and calls for use of a single shear fixture. For threaded fasteners the more common practice is to use a double shear fixture. At the present time there is no standard test method, although one is under development for the inclusion in ASTM F606. Unfortunately, variations in test fixture designs cause significant differences in measured shear strengths. Consequently when shear strength is a requirement an understanding of how it will be measured is advisable between the purchaser and his supplier.

As an empirical guide, shear strengths of fasteners may be assumed as being approximately 60% of their specified minimum tensile strength. For example, an SAE Grade 5 hex cap screw has a specified minimum tensile strength of 120,000 psi, consequently its shear strength for design purposes could safely be assumed to be 70,000 psi. When the shear plane occurs through the full body of the fastener, the shear area is computed using the nominal diameter. For the case of shearing through the threads, the shear area is the thread root area as given in Tables 1 and 2.

Torsional strength is a load usually expressed in terms of torque, at which a fastener fails by being twisted off about its axis. Tapping screws and stainless steel metric machine screws (ASTM F738) are the only industrial fasteners for which torsional strength is specified as a requirement. When conducting the torsional strength test, the screw is clamped in its threaded length in such a manner that rotation is prevented. The test setup also requires that two of its threads are exposed above the clamping device. The specified torsional strength torque is applied by tightening the head. The screw is assumed to pass the test if its head does not twist off.

Fatigue strength is the maximum stress a fastener can withstand for a specified number of repeated applications or cycles prior to its failure. Stress level and cycles to failure are inescapably related. Obviously, for a single load application the "fatigue strength" equals the static tensile strength. As the applied stress is reduced the component can endure an increasing number of load applications until it reaches the other extreme of its fatigue strength range known as its endurance limit. This limit corresponds to the stress the product can accept and survive indefinitely. While aerospace fasteners are frequently tested for their fatigue strength properties, industrial fasteners are practically never subjected to such tests. The principal reasons are testing difficulty and expense. But, perhaps of greater importance is the fact that the fatigue-resistant properties of the fastener rarely have a known relationship to the structural fatigue environment of the service application.

The preceding discussion of mechanical properties applies to externally threaded fasteners. Strength characteristics of internal threads warrant additional considerations.

14

Strength of Internal Threads

The single most important mechanical property is *proof load* which is a load, axially applied through a mating bolt or hardened mandrel, that the nut must support without any evidence of thread stripping or rupturing of the nut wall. Additionally, after the proof load has been applied and the load released, the nut must be freely removable from the test bolt or mandrel. This latter requirement is interesting because it demonstrates that insufficient thread distortion has occurred to freeze the mating threads together. Tests have shown that a distortion of this magnitude happens when the applied load reaches about 95% of the thread strip-out load.

In nut strength-grade standards, proof loads are usually specified as proof stresses in pounds per square inch (psi) for inch series and megapascals (MPa) for metric series nuts. To compute proof loads in pounds multiply the proof stress (psi) by the thread stress area as given in Table 1. For metric nuts multiply the proof stress (MPa) by the thread stress area (mm²) as given in Table 2 and divide by 1000 to get kilonewtons (kN). It should also be noted that nut proof loads are basically unrelated to proof loads specified for externally threaded fasteners.

Cone proof load is an axially applied load a nut must withstand when seated on a hardened cone shaped washer with an included angle of 120°.

59

The purpose of the cone-washer is to introduce a severe nut wall dilating action as the nut is axially stressed. Should the nut fail during such testing it will usually rupture by splitting rather than thread stripping. Nuts intended for high temperature service (ASTM A194) are cone-proof load tested. Also, standards defining acceptable limits for surface discontinuities on nuts require cone-proof load testing of nuts with seams to prove that the seam is of insufficient magnitude to deleteriously influence service performance. Cone-proof load stresses can be computed by multiplying the specified proof load of the nut by the factor (1-0.3D) for inch series nuts and by (1-0.12D) for the metric configuration where D is the nominal nut size in inches and millimeters respectively. Cone-proof load testing of nuts is not entirely definitive in identifying unacceptable nuts. However it is the best test procedure so far developed and it will continue to be a viable requirement for nuts until a more discriminating technique is introduced.

Hardness is an important mechanical property of nuts. Not only is hardness testing easy and quick to perform but, more importantly, it is the only known practical method of testing nuts that may have proof loads beyond the capability of available testing equipment. Generally, nuts with specified proof loads less than 120,000 lbs (530 kN) can be proof load tested. However, the nuts of higher strengths, which are beyond the testing capacity of the more common tensile testing machines, must be acceptance-tested on the basis of their meeting or exceeding the specified minimum hardness. Nut strength grade standards require that nuts of all sizes, grades and classes be hardness tested to assure their hardness does not exceed the specified maximum. This is especially important for heat treated nuts to demonstrate the nut has been properly processed to control potential brittleness.

15

Carbon Steel Fasteners

Over 90% of all fasteners being manufactured and used are made of carbon steel. And the reason is quite simple. Steel has excellent workability, it offers the broadest range of attainable combinations of strength properties, and it is inexpensive. There are well over 100 different strength grades for steel fasteners, each with its own set of properties and designation. All of the different grades serve a genuine need although the majority of them are special purpose. Fortunately there are only a relatively few grades with which the engineer needs to have more than a casual acquaintance.

In general, carbon steel fastener strength grades can be catalogued into three broad groupings involving low carbon, medium carbon, and alloy steel. The best known and the most widely referenced strength grade system for externally threaded fasteners is detailed in the SAE J429 standard. This comprehensive system comprises 10 separate specifications spanning low carbon steel Grade 1 through alloy steel Grade 8. The more important grades of the SAE system are repeated and expanded on in individual ASTM standards, principally A307, A449, A325, A354, and A490. The strength grade system for carbon steel metric fasteners is covered in ISO 898/I which closely parallels the content and structure of SAE J429. ASTM F568 is the domestic version of ISO 898/I, detailing those property classes of metric fasteners which are gaining popularity in North America.

For fasteners, *low carbon steels* are those with insufficient carbon to permit a predictable response to a strengthening heat treatment. The most commonly used analyses are AISI 1006, 1008, 1016, 1018, 1021, and 1022. These steels have excellent workability and good strength properties which can be further improved by cold working. They can also be case hardened and welded.

SAE Grade 1, ASTM A307 Grade A, and ASTM F568 class 4.6 are low carbon steel strength grades with essentially the same properties. A307 Grade B is a special low carbon steel grade of bolt used in piping and flange work. Its properties are identical with those of Grade A except that it has the added requirement of a specified maximum tensile strength. The reason for this is to make sure that if a bolt is inadvertently overtightened during installation it will fracture before breaking the cast iron flange of an expensive length of pipe, valve, or pump. SAE Grade 2 and F568 classes 4.8 and 5.8 are low carbon steel strength grades which recognize the strength improvement gained by the fastener through cold working.

Medium carbon steels are heat treatable. This means that through metallurgical treatment the tensile strength of the fastener can be made substantially higher than that of the raw material. The popular chemical analyses for these materials conform to AISI 1030, 1035, 1038, and 1541 standards. These steels have good workability. However as carbon content increases, heading difficulty increases while the life of the manufacturing tools and dies decreases. To compensate, the raw material is frequently annealed in a pretreatment process such as normalizing or spheroidizing.

Because the heat treatment response is due solely to carbon content without the booster effect of alloying elements, the end properties of the fastener are subject to size effect. This means that using the same material analysis and the same processing the strength properties of the fastener will decrease as its size increases. This is why SAE Grade 5 and ASTM A449 for inch series fasteners have their strength properties "stepped down" for the larger sizes over those specified for the smaller diameters. ISO property classes 8.8 and 9.8 partially recognize a similar pattern for metric fasteners by applying class 9.8 with its higher strength properties to products with sizes 16 mm and smaller and class 8.8 to larger size fasteners. The strength requirements of class 8.8 are readily attainable using medium carbon steel for fasteners as large as 24 mm, but for larger sizes it is frequently necessary to revert to an alloy steel with its superior heat treatment response to achieve the specified strength properties. And when use of alloy steel becomes necessary, the engineer might usefully ask why limit the properties to those specified for a medium carbon steel when normal processing will automatically give the higher strengths specified for alloys steel fasteners. This suggests that class 8.8 metric fasteners in sizes larger than 24 mm may not always be the most economical choice and that perhaps the engineer should consider class 10.9.

On a strength-to-cost basis, medium carbon heat treated fasteners provide more load carrying capability per unit of cost than any other known metal. Their yield-to-ultimate strength ratio is the lowest of all heat treated steels which assures their superior ductility. In fact, they are frequently referred to as "forgiving" which means they have a punching bag ability to absorb punishment and service abuse. Medium carbon steels embody the most attractive balance between cost, manufacturing convenience, and superlative mechanical properties. This is why SAE Grade 5, ASTM A449, ASTM A325, and F568 classes 8.8 and 9.8 collectively are the most popular strength-grades for externally threaded fasteners in use today.

Carbon steel is classed as an *alloy steel* when the maximum of the range of content specified for manganese is greater than 1.65%, for silicon 0.60%, for copper 0.60%, or when the chromium content is less than 4% (if greater it approaches being a stainless steel). Also, when the steel contains a specified minimum content of aluminum, boron, cobalt, columbium, molybdenum, nickel, titanium, vanadium, zirconium or any other element added to achieve specific effect we class carbon steel as an alloy steel. Literally dozens of different alloy steels are used in the manufacture of fasteners with the most popular standard analyses (and their principal alloying elements) being AISI 1335 (Mg), 4037 (Mo), 4140 (Cr Mo), 4340 (Ni Cr Mo), 8637 (Ni Cr Mo) and 8740 (Ni Cr Mo). Perhaps a look at the effects these alloying elements have on the performance of the steel might be useful in understanding their popularity. A brief overview of these characteristics is given in the next chapter.

16

Effects of Alloying Elements

Manganese contributes strength and moderately improves hardenability, which means it enhances the depth of hardness penetration after quenching. Manganese is generally beneficial to surface quality, however, high manganese content adversely affects ductility and weldability. Copper, when present in amounts exceeding 0.20%, enhances resistance to atmospheric corrosion. Nickel is a particularly valuable alloying element. It provides strength, improves toughness at low temperatures, benefits corrosion resistance, and adds a pacifying quality to the heat treatment process to assure more consistent and fool proof results. Chromium increases hardenability, improves wear and abrasion resistance, adds to corrosion resistance, and helps retain strength at high temperatures. Molybdenum helps control hardenability, has a powerful effect on increasing high temperature tensile and creep strengths, and reduces the susceptability of the steel to temper brittleness. Boron enhances hardenability and is most helpful in providing a predictable heat treatment response to low carbon steels. The combination of two or more alloying elements gives the steel the characteristic properties of each. In particular, the combined effect on hardenability is considerably greater than the sum of the same elements used individually.

In recent years low carbon martensitic steels, popularly known as boron treated steels, have gained a high degree of acceptance as alternate materials providing the strength properties expected of medium carbon heat treated steels and even higher. The low carbon content bypasses the need to pretreat (anneal) the material prior to heading. Formability is better, tool life is extended and surface quality improved. The popular alloys are 10B18, 10B21, and 10B22. These steels have sufficient hardenability to meet the strength requirements of SAE Grade 8 but the slightly carbon richer alloys, such as 10B30, are frequently used for higher strength demands. Boron steels are economically attractive, however they are not without their drawbacks. Because of their lower tempering temperature their stress relaxation properties at moderately elevated temperature are inferior. Also, the fastener manufacturer must exercise closer control during heat treatment. The margin for error is not quite as generous as when heat treating steels having richer alloying contents.

Another family of alloy steels is the group of atmospheric corrosion resistant steels with their copper-manganese-nickel-chromium alloying content. Fasteners of these materials have their greatest popularity in structural applications such as bridges, buildings, and towers because they can be installed without surface protection. The chemical compositions of these steels are proprietary and most are trade named.

SAE Grade 8, ASTM A354 Grade BD, A490, and F568 class 10.9 are alloy steel strength-grades for fasteners having essentially the same mechanical properties. SAE Grades 5.2 and 8.2, ASTM A325 Type 2 and A490 Type 2 are boron treated steels. Boron steels are also permitted as alternatives for classes 8.8, 9.8, and 10.9. Type 3 bolts of A325 and A490 are made of atmospheric corrosion resistant steel.

17

Fastener Grades

The mechanical properties and grade identification markings for the popular low carbon, medium carbon and alloy steel strength grades for externally threaded inch and metric series fasteners are detailed in Tables 3 through 7.

The most widely referenced strength grade system for inch series externally threaded fasteners throughout the world today is that of SAE. It is detailed in SAE Standard J429, comprising 10 different strength levels each identified by a number ranging from 1 through 8.2. The numbers have no particular significance except that increasing numbers represent increasing tensile strengths. The relevant decimals of whole numbers represent the same basic properties of the whole number but a variation in material and treatment. Of the 10 grades there are only six which have genuine popularity. These are covered in Tables 3, 4 and 5. The four omitted grades are 4, 5.1, 7, and 8.1. Grade 4 is for studs made of cold drawn medium carbon steel representing a minimum tensile strength of 115,000 psi. The cold working of the raw material bypasses the need to heat treat the studs after their manufacture. Grade 5.1 is a special provision for screw-washer assemblies (commercially known as sems) and other small screws of 5/8 in. size and smaller. Grade 5.1 products are made of low or medium carbon steel, are quenched and tempered, and have strength properties paralleling those of Grade 5. Grade 7 is a

Table 3   Mechanical Requirements for Carbon Steel Externally Threaded Fasteners, Inch Series

Grade Designation	Nominal size of Product (in.)[a]	Material and Treatment		Mechanical Requirements			Product hardness, Rockwell			Grade Identification Marking
			Proof load (stress) (ksi)	Yield strength (ksi) Min	Tensile strength (ksi) Min	Surface Max	Core Min	Core Max		
A307 Gr. A	1/4 to 4	Low or medium carbon steel	—	—	60	—	B69	B100	none
A307 Gr. B	1/4 to 4		—	—	60 min 100 max	—	B69	B95	specified
SAE Gr. 1	1/4 to 1 1/2	steel	33	36	60	—	B70	B100	none
SAE Gr. 2	1/4 to 3/4	low or medium carbon steel, cold worked	55	57	74	—	B80	B100	specified
SAE Gr. 5	1/4 to 1	medium carbon steel; the product is quenched and tempered	85	92	120	30N54	C25	C34	
	1 1/8 to 1 1/2		74	81	105	30N50	C19	C30	
A449	1/4 to 1		85	92	120	—	C25	C34	
	1 1/8 to 1 1/2		74	81	105	—	C19	C30	
	1 3/4 to 3		55	58	90	—	B91	B100	
A325 Type 1	1/2 to 1	and tempered	85	92	120	—	C24	C35	A325
	1 1/8 to 1 1/2	pered	74	81	105	—	C19	C31	
SAE Gr. 5.2	1/4 to 1	low carbon steel; is quenched and tempered	85	92	120	30N56	C26	C36	
A325 Type 2	1/2 to 1		85	92	120	—	C24	C35	
	1 1/8 to 1 1/2		74	81	105	—	C19	C31	A325

Grade	Size range	Material							Grade marking
A325 Type 3	1/2 to 1	atmospheric corrosion resistant steel; product is quenched and tempered	85	92	120	—	C24	C35	(marking)
	1 1/8 to 1 1/2		74	81	105	—	C19	C31	A325
SAE Gr. 8	1/4 to 1 1/2	medium carbon alloy steel; product is quenched and tempered	120	130	150	30N58.6	C33	C39	(marking)
A354 Gr. BD	1/4 to 2 1/2	product is quenched and tempered	120	130	150	—	C33	C39	(marking)
	2 3/4 to 4		105	115	140	—	C31	C38	
A490 Type 1	1/2 to 1 1/2		120	130	150 min 170 max	—	C33	C38	A490
SAE Gr. 8.2	1/4 to 1	low carbon boron steel; product is quenched and tempered	120	130	150	30N58.6	C33	C39	(marking)
A490 Type 2	1/2 to 1	product is quenched and tempered	120	130	150 min 170 max	—	C33	C38	A490
A490 Type 3	1/2 to 1 1/2	atmospheric corrosion resistant steel; product is quenched and tempered	120	130	150 min 170 max	—	C33	C38	A490

[a] SAE Grade 2 products are available only in lengths 6 in. and shorter.

Table 4 Proof Loads and Tensile Strengths of Carbon Steel Externally
Threaded Fasteners, Inch Series, Coarse Threads (All loads are in 1000
pounds)

Nominal Product Size and Threads (per in.)	SAE Grade 1 A307 Gr. A&B		SAE Grade 2		SAE Gr. 5&5.2 A449 A325 Types 1,2,3		SAE Gr. 8&8.2 A354 Gr. BD A490 Types 1,2,3	
	Proof Load	Tensile Strength Min	Proof Load	Tensile Strength Min	Proof Load	Tensile Strength Min	Proof Load	Tensile Strength Min
1/4-20	1.05	1.91	1.75	2.35	2.70	3.82	3.82	4.77
5/16-18	1.73	3.14	2.88	3.88	4.45	6.29	6.29	7.86
3/8-16	2.56	4.65	4.26	5.74	6.59	9.30	9.30	11.6
7/16-14	3.51	6.38	5.85	7.87	9.04	12.8	12.8	15.9
1/2-13	4.68	8.51	7.80	10.5	12.1	17.0	17.0	21.3
9/16-12	6.01	10.9	10.0	13.5	15.5	21.8	21.8	27.3
5/8-11	7.46	13.6	12.4	16.7	19.2	27.1	27.1	33.9
3/4-10	11.0	20.0	18.4	24.7	28.4	40.1	40.1	50.1
7/8-9	15.2	27.7	—	—	39.3	55.4	55.4	69.3
1 - 8	20.0	36.4	—	—	51.5	72.7	72.7	90.9
1 1/8-7	25.2	45.8	—	—	56.5	80.1	91.6	114
1 1/4-7	32.0	58.1	—	—	71.7	102	116	145
1 3/8-6	38.1	69.3	—	—	85.5	121	139	173
1 1/2-6	46.3	84.3	—	—	104	148	169	211
1 3/4-5	62.7	114	—	—	105	171	228	285
2 -4 1/2	82.5	150	—	—	138	225	300	375
2¼-4½	107	195	—	—	179	293	390	488
2½-4	132	240	—	—	220	360	480	600
2 3/4-4	163	296	—	—	271	444	518	690
3 - 4	197	358	—	—	328	537	629	836
3 1/4-4	234	426	—	—	—	—	746	994
3 1/2-4	275	500	—	—	—	—	875	1170
3 3/4-4	319	580	—	—	—	—	1010	1350
4 - 4	366	665	—	—	—	—	1160	1550

Proof loads and tensile strengths are computed by multiplying the respective stresses
given in Table 3 by the thread stress area given in Table 1.
Externally threaded fasteners of sizes and grades where no strength values are given
are nonstandard.

Table 5 Proof Loads and Tensile Strengths of Carbon Steel Externally
Threaded Fasteners, Inch Series, Fine Threads (All loads are in 1000 pounds)

Nominal Product Size and Threads (per in.)	SAE Grade 1 A307 Gr. A&B		SAE Grade 2		SAE Gr. 5&5.2 A449		SAE Gr. 8&8.2 A354 Gr. BD	
	Proof Load	Tensile Strength Min	Proof Load	Tensile Strength Min	Proof Load	Tensile Strength Min	Proof Load	Tensile Strength Min
1/4-28	1.20	2.18	2.00	2.69	3.09	4.37	4.37	5.46
5/16-24	1.91	3.48	3.19	4.29	4.93	6.96	6.96	8.70
3/8-24	2.90	5.27	4.83	6.50	7.46	10.5	10.5	13.2
7/16-20	3.92	7.12	6.53	8.78	10.1	14.2	14.2	17.8
1/2-20	5.28	9.59	8.79	11.8	13.6	19.2	19.2	24.0
9/16-18	6.70	12.2	11.2	15.0	17.3	24.4	24.4	30.5
5/8-18	8.45	15.4	14.1	18.9	21.8	30.7	30.7	38.4
3/4-16	12.3	22.4	20.5	27.6	31.7	44.8	44.8	56.0
7/8-14	16.8	30.5	–	–	43.3	61.1	61.1	76.4
1 - 12	21.9	39.8	–	–	56.4	79.6	79.6	99.5
1 - 14	22.4	40.7	–	–	57.7	81.5	81.5	102
1 1/8-12	28.2	51.4	–	–	63.3	89.9	103	128
1 1/4-12	35.4	64.4	–	–	79.4	113	129	161
1 3/8-12	43.4	78.9	–	–	97.3	138	158	197
1 1/2-12	52.2	94.9	–	–	117	166	190	237

Proof loads and tensile strengths are computed by multiplying the respective stresses given in Table 3 by the thread stress area given in Table 1.

Externally threaded fasteners of sizes and grades where no strength values are given are non-standard.

special grade for fasteners of alloy steel which are roll threaded after heat treatment to enhance their fatigue resistant properties. Grade 7 strength properties are modestly lower than those of Grade 8. However fasteners with Grade 8 properties which are roll threaded after heat treatment are also available. Such fasteners are quite expensive because of the lowered die life when roll threading material of a relatively high hardness. Only in the rarest applications can such an additional cost be justified. Grade 8.1 is limited to studs made of 1541 steel which had been drawn at a slightly elevated temperature to gain the high strength properties without the need to heat treat the stud after its manufacture. Grade 8.1 strength properties are the same as those of Grade 8.

ASTM standard has no single integrated system presented in one document but rather has a number of separate standards each covering the requirements for a single grade. The principal specifications are A307, A449, and A354 and these essentially parallel SAE grades 1, 5, and 8. One of the advantages of the ASTM standards is that they cover product sizes through 4 in. The largest size recognized in SAE J429 is 1-1/2 in. ASTM A325 and A490 are two standards which are frequently misapplied. Both are limited exclusively to high strength structural bolts in sizes 1/2 through 1-1/2 in. They do not cover such products as anchor bolts, studs, or carriage bolts. Properties of these special products are covered in A449 and A354 specifications.

Four ASTM standards not discussed in this book in great detail are A574, A687, A193, and A320. A574 covers alloy steel socket head cap screws in sizes No. 0 through 4 in. These screws have a tensile strength of 180,000 psi for sizes 1/2 in. and smaller. Larger screws are associated with 170,000 psi strength. A574 is not a strength grade available for use with any other style of externally threaded fastener since it was written strictly for socket head cap screws. The A687 standard covers alloy steel studs and non-headed bolts intended for use as anchor bolts. A687 products have improved impact strength properties at low temperatures in addition to their specified minimum tensile strength of 150,000 psi. The A193 standard pertains to several grades of fasteners suitable for elevated temperature applications with the A320 specification relating to low temperatures. While each recognizes some alloy steel grades, most of their grades are austenitic stainless steel.

In the popular SAE and ASTM strength grade systems the highest strength grade recognized (with the exception of A574) are fasteners with a specified minimum tensile strength of 150,000 psi. From time to time consideration is given to expanding the systems to add a higher strength material. However the responsible technical committees have so far rejected such proposals. The compelling reason for this philosophy is that designers could unwittingly specify use of the higher strength grade without a full understanding of the related risks. For instance, fasteners with tensile strengths exceeding 160,000 psi have hardness ranges exceeding Rockwell C40. Extensive research, coupled with highly unfortunate field experience, has shown that unless exceptional precautionary steps are taken during manufacture, products of hardnesses higher than C39 have an unacceptably high susceptibility to stress embrittlement. The higher the specified tensile strength, the higher the hardness and the more critical becomes the choice of the material analysis and the process of heat treatment. Producers of aerospace fasteners and socket screws have the necessary experience and skills in these matters. Unfortunately, many others do not.

Metric strength grades are called *property classes*. These terms originated in ISO standards and were copied into SAE and ASTM specifications. The ISO property class system for externally threaded metric fasteners is presented in ISO 898/1. Seven of its 10 classes should have extensive use by North American industry and are covered in the SAE J1199 and ASTM F568 standards. Both are in essential agreement with ISO 898/1, although F568 is more complete because it recognizes fastener sizes through 100 mm while 898/1 and J1199 limit their coverage to 36 mm. F568 also includes two classes for structural bolts made of atmospheric corrosion resistant steel.

Property class designations are numerals and they are quite significant. The numeral(s) preceding the first decimal point approximates 1/100 of the minimum specified tensile strength in MPa. The numeral following the first decimal point approximates 1/10 of the ratio expressed as a percentage between the minimum yield strength and the minimum tensile strength. The numeral 3 following the second decimal point indicates the material is an atmospheric corrosion resistant steel. As an example, class 4.6 has a specified minimum tensile strength of 400 MPa (4 × 100) and a minimum yield strength of 240 MPa (0.6 × 400). Not all of the designations give exact tensile and yield values in this way but each gives a reasonably close approximation. The properties of the metric property classes are detailed in Tables 6 and 7. In those instances where only the class number is given, the class is recognized in ISO, SAE, and ASTM standards and the indicated values are the same in each.

When designing in metric, it is strongly recommended that the engineer gives his first priority to fasteners of class 10.9 and lower strengths. It is true that the ISO system recognizes the higher strength 12.9 and that this class is repeated in both J1199 and F568. However its commercial use in North America at this time is limited to socket head cap screws. It is worth emphasizing again that fasteners with hardnesses exceeding Rockwell C39, which class 12.9 does, require very special care in their manufacture. Their use should not be specified casually. For designers who may not fully appreciate the attendant risk in using fasteners at these strength levels, the price for the modest strength improvement over class 10.9 can be high indeed.

It is a mandatory requirement of SAE and ASTM standards that inch series fasteners of the medium carbon and alloy steel strength grades and metric fasteners of all steel property classes be marked for grade identification. The

Table 6 Mechanical Requirements for Carbon Steel Externally Threaded Fasteners, Metric Series

Property Class Designation (a)	Nominal Size of Product	Material and Treatment	Proof Load Stress (MPa) Min	Yield Strength (MPa) Min	Tensile Strength (MPa) Min	Product Hardness Rockwell			Property Class Identification Marking
						Surface Max	Core Min	Core Max	
4.6	M5 to M100	low or medium carbon steel	225	240	400	–	B67	B95	4.6
4.8	M1.6 to M16	low or medium carbon steel, fully or partially annealed	310	340	420	–	B71	B95	4.8
5.8	M5 to M24 (b)	low or medium carbon steel, cold worked	380	420	520	–	B82	B95	5.8
8.8	M16 to M72	medium carbon steel; the product is quenched and tempered	600	660	830	30N56	C23	C34	8.8
A325M Type 1	M16 to M36	low carbon boron steel; the product is quenched and tempered	600	660	830	30N56	C23	C34	A325M 8S
8.8									8.8
A325M Type 2	M16 to M36	steel; the product is quenched and tempered	600	660	830	30N56	C23	C34	A325M 8S
A325M Type 3	M16 to M36	atmospheric corrosion resistant steel; the product is quenched and tempered	600	660	830	30N56	C23	C34	A325M 8S3
9.8	M1.6 to M16	medium carbon steel the product is quenched and tempered	650	720	900	30N58	C27	C36	9.8

Property class	Size	Material and condition						
9.8	M1.6 to M16	low carbon boron steel; the product is quenched and tempered	650	720	900	30N58	C27	C36 9.8
10.9	M5 to M20	medium carbon steel; the product is quenched and tempered	830	940	1040	30N59	C33	C39 10.9
10.9	M5 to M100	medium carbon alloy steel; the product is quenched and tempered	830	940	1040	30N59	C33	C39 10.9
A490M Type 1	M12 to M36							A490M 10S
10.9	M5 to M36	low carbon boron steel; the product is quenched and tempered	830	940	1040	30N59	C33	C39 10.9
A490M Type 2	M12 to M36							A490M 10S
A490M Type 3	M12 to M36	atmospheric corrosion resistant steel; the product is quenched and tempered	830	940	1040	30N59	C33	C39 A490M 10S3
12.9 (c)	M1.6 to M100	alloy steel; the product is quenched and tempered	970	1100	1220	30N63	C38	C44 12.9

a. When only the property class number is shown, the class is standard in ISO 898/1, ASTM F568, and SAE J1199. The properties specified in the three documents are identical with one or two minor exceptions. Where differences exist, the values given in F568 are shown.

b. Class 5.8 products are available only in lengths 150 mm and shorter.

c. Caution is advised when considering the use of class 12.9 products. Capability of the fastener manufacturer, as well as the anticipated in-use environment, should be considered. High strength class 12.9 products require rigid control of heat treating operations and careful monitoring of as-quenched hardness, surface discontinuities, depth of partial decarburization, and freedom from carburization. Some environments may cause stress corrosion cracking of nonplated as well as electroplated products.

Table 7 Proof Loads and Tensile Strengths of Carbon Steel Externally Threaded Fasteners, Metric Series (All loads are in kilonewtons (kN))

Nominal Product Size and Thread Pitch	Property Class													
	4.5		4.8		5.8		8.8 A325M Types 1,2,3		9.8		10.9 A490M Types 1,2,3		12.9	
	Proof Load	Tensile Strength Min	Proof Load	Tensile Strength Min	Proof Load	Tensile Strength Min	Proof Load	Tensile Strength Min	Proof Load	Tensile Strength Min	Proof Load	Tensile Strength Min	Proof Load	Tensile Strength Min
M1.6 X 0.35	—	—	0.39	0.53	—	—	—	—	0.38	1.14	—	—	1.23	1.55
M2 X 0.4	—	—	0.64	0.87	—	—	—	—	1.35	1.86	—	—	2.01	2.53
M2.5 X 0.45	—	—	1.05	1.42	—	—	—	—	2.20	3.05	—	—	3.29	4.14
M3 X 0.5	—	—	1.56	2.11	—	—	—	—	3.27	4.53	—	—	4.88	6.14
M3.5 X 0.6	—	—	2.10	2.85	—	—	—	—	4.41	6.10	—	—	6.58	8.27
M4 X 0.7	—	—	2.72	3.69	—	—	—	—	5.71	7.90	—	—	8.52	10.7
M5 X 0.8	3.20	5.68	4.40	5.96	5.40	7.38	—	—	9.23	12.8	11.8	14.8	13.8	17.3
M6 X 1	4.52	8.04	6.23	8.44	7.64	10.5	—	—	13.1	18.1	16.7	20.9	19.5	24.5
M8 X 1.25	8.24	14.6	11.3	15.4	13.9	19.0	—	—	23.8	32.9	30.4	38.1	35.5	44.7
M10 X 1.5	13.1	23.2	18.0	24.4	22.0	30.2	—	—	37.7	52.2	48.1	60.3	56.3	70.8
M12 X 1.75	19.0	33.7	26.1	35.4	32.0	43.8	—	—	54.8	75.9	70.0	87.7	81.8	103

Size														
M14 × 2	25.9	46.0	35.7	48.3	43.7	59.3	–	–	74.8	104	95.5	120	112	140
M16 × 2	35.3	62.8	48.7	65.9	59.7	81.6	94.2	130	102	141	130	163	152	192
M20 × 2.5	55.1	98.0	–	–	92.1	127	147	203	–	–	203	255	238	299
M22 × 2.5	–	–	–	–	–	–	182	251	–	–	251	315	–	–
M24 × 3	79.4	141	–	–	134	184	212	293	–	–	293	367	342	431
M27 × 3	–	–	–	–	–	–	275	381	–	–	381	477	–	–
M30 × 3.5	126	224	–	–	–	–	337	466	–	–	466	583	544	684
M36 × 4	184	327	–	–	–	–	490	678	–	–	678	850	792	997
M42 × 4.5	252	448	–	–	–	–	672	930	–	–	930	1160	1090	1370
M48 × 5	331	588	–	–	–	–	882	1220	–	–	1220	1530	1430	1790
M56 × 5.5	457	812	–	–	–	–	1220	1680	–	–	1680	2110	1970	2480
M64 × 6	603	1070	–	–	–	–	1610	2220	–	–	2220	2790	2600	3270
M72 × 6	779	1380	–	–	–	–	2080	2870	–	–	2870	3600	3360	4220
M80 × 6	977	1740	–	–	–	–	–	–	–	–	3600	4510	4210	5290
M90 × 6	1260	2240	–	–	–	–	–	–	–	–	4640	5810	5420	6820
M100 × 6	1570	2800	–	–	–	–	–	–	–	–	5800	7270	6780	8530

Proof loads and tensile strengths are computed by multiplying the respective stresses given in Table 6 by the thread stress areas given in Table 2 and dividing by 1000 to express the load in kilonewtons.

Externally threaded fasteners of sizes and property classes where no strength values are given are nonstandard.

M22 and M27 sizes are standard only for A325M and A490M high strength structural bolts.

77

only exceptions are slotted and recessed head screws and fasteners smaller than 5 mm. Additionally, and most importantly, the same standards require that all steel fasteners be further marked to identify the manufacturer. Without question one of the best guarantees of product quality is this requirement that the manufacturer identify himself as the producer of the part. Traceability, accompanying potential liability in case of a service failure, is ample incentive to any reputable producer to exercise all of the care necessary to manufacture fully conforming parts.

The data in Tables 3 through 7 is indispensable to any designer using steel fasteners. As will be expanded on later, this information coupled with that given on steel nuts in Tables 8 through 11 will provide the engineer with all that is really necessary to accomplish a reasonably respectable and safe design in most routine service applications.

18

Design Features of Nuts

Strengths of bolts and screws are solely dependent on their size, thread series, and material. The configuration of their head is unimportant providing, of course, it is strong enough to support the load. And only in the rarest of special head designs would there be a doubt.

However, strengths of nuts, while equally dependent on size, thread series, and material, are inescapably linked to their external geometry. Nut heights and wall thicknesses relate directly and significantly to load carrying capability. A nut's height establishes the length of thread engagement and its wall thickness establishes its resistance to dilation defined as a radial spreading out of the nut at its bearing face when it is axially stressed. Quite obviously, all other strength factors being equal, a thicker nut will support more load as will one with a greater width across its wrenching flats or a flanged bearing face. Consequently, nut strength grade systems must not only define mechanical properties but they must also identify the dimensional nut design to which the particular strength values apply.

For inch series nuts there are three basic dimensional designs. Regular hex nuts have formula dimensions of 0.875 D (where D is the nut's nominal size) for their thickness and 1.5 D for their width across flats. Thick hex nuts have the same widths across flats and an increased thickness equal to about 1 D.

Heavy hex nuts also have thicknesses of about 1 D and larger widths across flats equal to 1.5 D plus 1/8 in. Slotted nuts are standard for each of the foregoing three basic designs of hex nuts. The loss of thread engagement due to the slotting reduces the strength of the full form nut by about 20%. Hex and heavy hex jam nuts have thicknesses of about two-thirds those of full form nuts and their strengths may be assumed to be 60% of those for full thickness nuts of the same width across flats. Square nuts are still commercially available although their use has been discouraged for several years. Square nuts are normally produced only in the lowest strength grade.

The inch series designs of nuts, both dimensionally and strength-wise have been standard for well over 25 years. They have functioned reasonably well although during that time occasional problems surfaced and adjustments were necessary. It is doubtful if any further revisions will be made, certainly none are anticipated at this time.

By 1970 it was increasingly apparent that in the not too distant future North American industry would be predominately metric. A technical committee of US and Canadian fastener engineers was formed and assigned the awesome task of developing all of the engineering standards for the metric fasteners that industry would eventually need. One of its early decisions was to use this timely opportunity provided by pending metric conversion to, at long last, design nuts for their optimum performance when mated with the various strength grades of externally threaded fasteners. Several countries combined their research resources and working together in ISO technical committees designed the ISO Metric nut strength grade system. Properly used, the system guarantees as fool-proof safe bolt and nut assemblies as sound engineering and prudent economics can provide. As stated previously the basic design premise is that, even under the most adverse combination of factors and during an inadvertent overtightening of the assembly the bolt will break at least once in every 10 installations. Once installed, the failure mode if the assembly is overstressed will normally be bolt fracture and not thread stripping.

There are six dimensional designs of metric nuts tied specifically to certain property classes. These involve hex style 1, hex style 2, hex slotted, hex jam, heavy hex, and hex flange type. Hex style 1 nuts are available in sizes M1.6 through M36 and only in classes 5 and 10. Their widths across flats and thicknesses are not formula values but approximate 1.5 D and 0.88 D, respectively. Hex style 2 nuts are exclusively for classes 9 and 12 in sizes M3 through M36. They have the same widths across flats as style 1 nuts but are slightly thicker approximating 1 D. Hex slotted nuts are available in sizes M5 through M36 and in property classes 5 and 10. Their external dimensions are the same as those of hex style 2 nuts. However due to the loss of thread engagement by slotting their strengths are only 80% of those specified for classes 5 and 10 nuts. Hex jam nuts are standard in sizes M5 through M36 and in classes 04 and 05. Their widths across flats are the same as those for hex style 1 nuts but their thickness is only about 0.5 D. Heavy hex nuts are available in sizes M12 through M100 and in fact they are the only standard metric nuts in sizes larger than M36. Consequently, heavy hex nuts are available in several property classes.

Heavy hex nuts follow the same width across flats series as other hex nuts except that for each size its width across flats is the next larger. Their thicknesses approximate 1 D. Heavy hex nuts were designed similarly as other metric nuts except that their design is based on the premise that when during mass assembly the bolt/nut combination is overtightened the bolt will fracture at least nine out of every ten installations. Hex flange nuts are avilable in sizes M5 through M20 and only in classes 9 and 10.

19

Grades of Nut Strength

The strength of a nut is defined as its proof load which, as explained earlier, corresponds to the axially applied load the nut must support without any evidence of failure. The actual load necessary to strip the nut threads or rupture the walls will always be at least 5% greater than the proof load.

The most comprehensive strength grade system for carbon steel inch series nuts is presented in ASTM A563. It comprises eight different grades and covers nut sizes 1/4 in. through 4 in. SAE J995 is another standard covering nut strength grades but its coverage is limited to just the three most popular grades with the sizes of 1-1/2 in. and smaller. ASTM A194 is a specification for nuts intended for high temperature service but it has within its system two grades of nuts which have extensive use in commercial and industrial applications. Tables 8 and 9 present the proof load and hardness requirements for the most commonly specified styles and grades of inch series nuts.

The ISO metric nut strength system is described in ISO 898/2. It comprises nine separate property classes; however, only six of these are popularly used in North America. ASTM A563M covers these six classes and its requirements, and with only one or two minor exceptions, are in exact agreement with 898/2. Additionally, A563M goes considerably beyond the ISO system. It extends the size range beyond M36 to give strength requirements for the

Table 8 Mechanical Requirements for Carbon Steel Nuts, Inch Series With UNC, 8 UN, 6 UN, and Coarser Pitch Threads

Style of Nut and Dimensional Standard	Hex Nuts[a] ANSI B18.2.2			Hex Thick Nuts—ANSI B18.2.2[b] Heavy Hex Nuts—ANSI B18.2.2					
Strength Grade Designation	A563 Gr. A and SAE Gr. 2[c]	A563 Gr. B and SAE Gr. 5[c]	A563 Gr. DH and SAE Gr. B[d]	A563 Gr. A[c] and SAE Gr. 5[c]	A563 Gr. B and SAE Gr. C3[d,e]	A563 Gr. C and A194 Gr. 2[d]	A563 Gr. D and A194 Gr. 2d[d] SAE Gr. B[d]	A563 Gr. DH and A194 Gr. 2H[d]	A563 Gr. DH3 and A194 Gr. 2H[d]
Proof Stress psi	90,000	120,000[f] 105,000	150,000	100,000	133,000[f] 116,000	144,000	150,000	165,000	175,000
Hardness Rockwell, min/max	B68/C32	B69/C32	C24/C38[g]	B68/C32	B69/C32	B78/C32	B84/C38	C38 max	C24/C38
Nominal Nut Size and Threads per Inch	Proof Load, 1000 lbs (Values are for nuts with UNC threads[c,h,j])								
1/4-20	2.86	3.82	4.77	3.18	4.23	—	4.77	5.25	5.57
5/16-18	4.72	6.29	7.86	5.24	6.97	—	7.86	8.65	9.17
3/8-16	6.98	9.30	11.6	7.75	10.3	—	11.6	12.8	13.6
7/16-14	9.57	12.8	15.9	10.6	14.1	—	15.9	17.5	18.6
1/2-13	12	17.0	21.3	14.2	18.9	20.4	21.3	23.4	24.8
9/16-12	16.4	21.8	27.3	18.2	24.2	26.2	27.3	30.0	31.9
5/8-11	20.3	27.1	33.9	22.6	30.1	32.5	33.9	37.3	39.6
3/4-10	30.1	40.1	50.1	33.4	44.4	48.1	50.1	55.1	58.5
7/8-9	41.6	55.4	69.3	46.2	61.4	66.5	69.3	76.2	80.9
1-8	54.5	72.7	90.9	60.6	80.5	87.3	90.9	100	106
1 1/8-7	68.7	80.1	114	76.3	88.5	110	114	126	134
1 1/4-7	87.2	102	145	96.9	112	140	145	160	170
1 3/8-6	104	121	173	116	134	166	173	191	202

Size									
1 1/2-6	126	148	211	141	163	202	211	232	246
1 3/4-5	171	200	285	190	–	–	285	–	333
2 - 41/2	225	263	375	250	–	–	375	–	438
2 1/4-4 1/2	–	–	–	325	–	–	488	–	569
2 1/2-4	–	–	–	400	–	–	600	–	700
2 3/4-4	–	–	–	493	–	–	740	–	863
3 - 4	–	–	–	597	–	–	896	–	1045
3 1/4-4	–	–	–	710	–	–	1065	–	1242
3 1/2-4	–	–	–	833	–	–	1250	–	1460
3 3/4-4	–	–	–	966	–	–	1450	–	1690
4 - 4	–	–	–	1110	–	–	1660	–	1940

[a]Hex nuts are standard only in sizes 1-1/2 in. and smaller; however hex nuts in sizes to 2 in. are commercially available, consequently their proof load values are included for information.

[b]Hex thick nuts are normally available in sizes 1-1/2 in. and smaller and only in A563 Grades A, B, and DH and SAE Grades 5 and 8.

[c]The tabulated proof loads are for nuts which have not been overtapped to accomodate an excessive coating or plating thickness on the mating externally threaded fastener. When nuts are overtapped to provide assembdability with fasteners having heavy coatings or platings, such as those which are hot dip or mechanically galvanized, the tabulated proof load values shall be reduced by multiplying the proof load by the factor 0.75.

[d]The tabulated proof load values apply to all nuts whether overtapped or not.

[e]A563 Grades C and C3 are heavy hex nuts intended for use with A325 high strength structural bolts.

[f]The higher proof load stress applies to nuts in sizes 1 in. and smaller. The lower proof load stress applies to nuts larger than 1 in.

[g]The hardness range shown applies to A563 Grade DH nuts. SAE Grade 8 nuts have specified hardnesses of Rockwell C24/32 for sizes 5/8 in. and smaller, C26/34 for sizes 3/4 to 1 in., and C26/36 for sizes larger than 1 in.

[h]Proof loads for nuts with 8 UN, 6 UN, or coarser pitch threads may be computed by multiplying the proof stress by the thread stress area using the following formula for stress area given previously as a base for developing Tables 1 and 2:

$$AS = 0.7854 \left(D - \frac{0.9743}{n} \right)^2$$

where AS = stress area, sq. in.; D = nominal thread size, in.; n = number of threads per inch.

[i]The tabulated proof loads are for full thickness nuts. Proof loads of slotted nuts may be assumed to be 80% of those for full thickness nuts and proof loads for jam nuts may be assumed to be 60% of those for full thickness nuts.

Table 9 Mechanical Requirements for Carbon Steel Nuts, Inch Series with UNF, 12 UN and Finer Pitch Threads

Style of Nut and Dimensional Standard	Hex Nuts—ANSI B18.2.2				Hex Thick Nuts—ANSI B18.2.2 Heavy Hex Nuts—ANSI B18.2.2			
Strength Grade Designation	A563 Gr. A[a] / SAE Gr. 2[b]	A563 Gr. B / SAE Gr. 5[a]	A563 Gr. DH / SAE Gr. 8[b]		A563 Gr. A[a] / SAE Gr. 5[a]	A563 Gr. B / SAE Gr. 5[a]	A563 Gr. D / SAE Gr. 8[b]	A563 Gr. DH[b]
Proof Stress psi	80,000	90,000	109,000 / 94,000[c]	150,000	90,000	120,000 / 105,000[c]	150,000	175,000
Hardness Rockwell, min/max	B68/C32	C32 max	B69/C32	C24/C38[d]	B68/C32	B69/C32	B84/C38	C24/C38
Nominal Nut Size and Threads per inch	Proof Load, 1000 lbs (Values are for nuts with UNF threads[a,e,f])							
1/4-28	2.91	3.28	3.97	5.46	3.28	4.37	5.46	6.37
5/16-24	4.64	5.22	6.32	8.70	5.22	6.96	8.70	10.2
3/8-24	7.02	7.90	9.57	13.2	7.90	10.5	13.2	15.4
7/16-20	9.50	10.7	12.9	17.8	10.7	14.2	17.8	20.8
1/2-20	12.8	14.4	17.4	24.0	14.4	19.2	24.0	28.0
9/16-18	16.2	18.3	22.1	30.5	18.3	24.4	30.5	35.5
5/8-18	20.5	23.0	27.9	38.4	23.0	30.7	38.4	44.8
3/4-16	29.8	33.6	40.7	56.0	33.6	44.8	56.0	65.3

Size	40.7	45.8	55.5	76.4	45.8	61.1	76.4	89.1
7/8-14								
1 - 12	53.0	59.7	72.3	99.5	59.7	79.6	99.5	116
1 - 14	54.4	61.2	74.1	102	61.2	81.6	102	119
1 1/8-12	68.4	77.0	80.5	128	77.0	89.9	128	150
1 1/4-12	85.8	96.6	101	161	96.6	113	161	188
1 3/8-12	105	118	124	197	118	138	197	230
1 1/2-12	126	142	149	237	142	166	237	277

[a] The tabulated proof loads are for nuts which have not been overtapped to accommodate excessive coating or plating thickness on the mating externally threaded fastener. When nuts are overtapped to provide assemblability with fasteners having heavy coatings or platings, such as those which are hot dip or mechanically galvanized, the tabulated proof load values shall be reduced by multiplying the proof load by the factor 0.75.

[b] The tabulated values apply to all nuts of the grade whether overtapped or not.

[c] The higher proof load stress applies to nuts in sizes 1 in. and smaller. The lower proof load stress applies to nuts larger than 1 in.

[d] The hardness range shown applies to A563 Grade DH nuts. SAE Grade 8 nuts have specified hardnesses of Rockwell C24/32 for sizes 5/8 in. and smaller, C26/34 for sizes 3/4 to 1 in., and C26/36 for sizes larger than 1 in.

[e] Proof loads for nuts with 12 UN and finer pitch threads may be computed by multiplying the proof stress by the thread stress area using the following formula for stress area (also quoted in notes for Table 8)

$$AS = 0.7854 \left(D - \frac{0.9743}{n} \right)^2$$

where: AS = stress area, sq. in.; D = nominal thread size, in.; n = number of threads per inch.

[f] The tabulated proof loads are for full thickness nuts. Proof loads of slotted nuts may be assumed to be 80% of those for full thickness nuts and proof loads for jam nuts may be assumed to be 60% of those for full thickness nuts.

Table 10 Mechanical Requirements for Carbon Steel Metric Nuts (Coarse Thread Series)[a]

Nominal Nut Size	Style of Nut[b]	Proof Load Stress MPa	Hardness Rockwell min/max
Class 04			
M5 to M36	Hex Jam	380	B89/C30
Class 5[c]			
M1-6 to M4		520	
M5 and M6	Hex	580	
M8 and M10	Style 1	590	B70/C30
M12 to M16		610	
M20 to M36		630	B78/C30
M42 to M100	Heavy Hex	630	B70/C30
Class 9			
M3 to M4	Hex Style 2	900	B85/C30
M5 and M6	Hex Style 2	915	
M8 and M10	and	940	
M12 to M16	Hex Flange	950	B89/C30
M20		920	
M24 to M36	Hex Style 2	920	
M42 to M100	Heavy Hex	920	
Class 12			
M3 to M6		1150	
M 8 and M10	Hex	1160	
M12 to M16	Style 2	1190	C26/36
M20 to M36		1200	
M42 to M100	Heavy Hex	1200	
Classes 8S and 8S3			
M12 to M36	Heavy Hex	1075	B89/C38
Class 05			
M5 to M36	Hex Jam	500	C26/C36
Class 5 Overtapped			
M5 and M6		465	
M8 and M10	Hex	470	B70/C30
M12 to M16	Style 1	490	
M20 to M36		500	B78/C30
M42 to M100	Heavy Hex	500	B70/C30

Table 10   (Continued)

Nominal Nut Size	Style of Nut[b]	Proof Load Stress MPa	Hardness Rockwell min/max
Class 10[c]			
M1-6 to M4	Hex Style 1	1040	
M5 to M10	Hex Style 1	1040	
M12 to M16	and	1050	C26/C36
M20	Hex Flange	1060	
M24 to M36	Hex Style 1	1060	
Class 12 Overtapped			
M5 and M6		920	
M8 and M10	Hex	930	
M12 and M16	Style 2	950	C26/C36
M20 to M36		960	
M42 to M100	Heavy Hex	960	
Classes 10S and 10S3			
M12 to M36	Heavy Hex	1245	C26/C38
Class 10 Overtapped			
M12 to M36	Heavy Hex	1165	C26/C38

[a]All values are as given in ASTM A563M. Values for classes 04, 05, 5, 9, 10, and 12 non-overtapped nuts in sizes M36 and smaller are identical to those given in ISO 898/2.
[b]Dimensions for hex style 1 nuts are given in ANSI B18.2.4.1M; for hex style 2 nuts in ANSI B18.2.4.2M; for hex flange nuts in ANSI B18.2.4.4M; for hex jam nuts in ANSI B18.2.4.5M; and for heavy hex nuts in ANSI B18.2.4.6M.
[c]Classes 5 and 10 hex slotted nuts with dimensions conforming to ANSI B18.2.4.3M are available. Proof load stresses are 80% of those shown for classes 5 and 10 hex style 1 nuts respectively.

larger sizes of nuts through M100. This specification recognizes three classes of nuts which have threads overtapped to provide easy assembly with galvanized bolts. It also includes four classes of nuts designed specifically for use with metric high strength structural bolts. A563M is the most comprehensive of any of the published standards giving metric nut strength data and its use is recommended. Tables 10 and 11 present the mechanical requirements for metric nuts.

Table 11 Proof loads of Metric Nuts, in kN

Nominal Nut Size and Thread Pitch	Property Class of Nut										
	04	05	5	5 (overtapped)	9	10	12	12 (overtapped)	8S and 8S3	10S and 10S3	10S (overtapped)
M1.6 X 0.35	—	—	0.66	—	—	1.32	—	—	—	—	—
M2 X 0.4	—	—	1.08	—	—	2.15	—	—	—	—	—
M2.5 X 0.45	—	—	1.76	—	—	3.23	—	—	—	—	—
M3 X 0.5	—	—	2.62	—	4.53	5.23	—	—	—	—	—
M3.5 X 0.6	—	—	3.53	—	6.10	7.05	—	—	—	—	—
M4 X 0.7	—	—	4.57	—	7.90	9.13	—	—	—	—	—
M5 X 0.8	5.40	7.1	8.24	6.60	13.0	14.8	16.3	13.1	—	—	—
M6 X 1	7.64	10.1	11.7	9.35	18.4	20.9	23.1	—	—	—	—
M8 X 1.25	13.9	18.3	21.6	17.2	34.4	38.1	42.5	—	—	—	—
M10 X 1.5	22.0	29.0	34.2	27.3	54.5	60.3	67.3	—	—	—	—
M12 X 1.75	32.0	42.2	51.4	41.3	80.1	88.5	100	—	90.6	105	98.2
M14 X 2	43.7	57.5	70.2	56.4	109	121	137	—	124	143	134
M16 X 2	59.7	78.5	95.8	76.9	149	165	187	—	169	195	183

Size											
M20 X 2.5	93.1	123	154	123	225	260	294	—	263	305	285
M22 X 2.5	—	—	—	—	—	—	—	—	326	377	353
M24 X 3	134	177	222	177	325	374	424	—	379	439	411
M27 X 3	—	—	—	—	—	—	—	—	493	571	535
M30 X 3.5	213	281	353	281	516	595	673	—	603	698	654
M36 X 4	310	409	515	409	752	866	980	784	878	1020	952
M42 X 4.5	—	—	706	560	1030	—	1340	1080	—	—	—
M48 X 5	—	—	920	735	1350	—	1760	1410	—	—	—
M56 X 5.5	—	—	1280	1020	1870	—	2440	1950	—	—	—
M64 X 6	—	—	1690	1340	2470	—	3220	2570	—	—	—
M72 X 6	—	—	2180	1730	3180	—	4150	3320	—	—	—
M80 X 6	—	—	2730	2170	3990	—	5210	4170	—	—	—
M90 X 6	—	—	3520	2800	5140	—	6710	5370	—	—	—
M100 X 6	—	—	4400	3500	6430	—	8390	6710	—	—	—

Proof loads are computed by multiplying the proof load stress (Table 10) by the thread stress areas (Table 2) and dividing by 1000 to express the load in kilonewtons.

Nuts of sizes and property classes where proof loads are shown are only standard in the dimensional style(s) indicated for the appropriate size and property class. Nuts of sizes and property classes where no proof load is given are nonstandard.

Nut sizes M22 and M27 nuts are structural nuts intended for use with high strength structural bolts. These sizes are available only in heavy hex nuts and of the property classes indicated.

A563 Grade A, SAE Grade 2, and A563M class 5 nuts are made of low carbon steel, usually a resulphurized grade such as AISI 1110 convenient for thread tapping. A563 Grades B and C, SAE Grade 5, and A563 M classes 04, 9, and 8S nuts may be made of low carbon steel such as 1018 or even one of the resulphurized grades. At times however, because of nut size or thread series, it is necessary to heat treat the nut to meet its specified strength requirements. Consequently a medium carbon steel such as 1038 is recommended. A563 Grades D and DH, SAE Grade 8, and A563M classes 05, 10, 10S, and 12 are made of medium carbon steel followed by a heat treatment. A194 Grades 2 and 2H are heavy hex nuts fabricated from steel with a minimum carbon content of 0.40%. The typical steel here is 1045. Grade 2 does not need to be heat treated, Grade 2H must. Currently Grade 2H nuts are the strongest carbon steel nuts commercially available. A563 Grades C3 and DH3, as well as A563M classes 8S3 and 10S3, are made of atmospheric corrosion resistant steel. Grade C3 and class 8S3 nuts are not heat treated; Grade DH3 and class 10S3 nuts are.

20

Overtapping and Identity
of Grades

When hot-dip or mechanically galvanized fasteners are used, or for that matter any coated or plated fastener with a coating thickness that consumes and
exceeds the thread allowance, some adjustment must be made in the threads
before the bolt and nut can assemble. North American practice is to always
overtap the nut thread and never undercut the bolt thread. The process of
overtapping usually follows the plating or coating operation.

Overtapping practice for inch series nuts is rather loosely controlled. The
standards only specify a minimum amount of overtapping. However the
overtapped nut needs to meet specified strength values representing a reasonable control of excessive overtapping. Nevertheless, for years there has been
a continuing—and festering—complaint. When fasteners with heavy coatings
or platings are used, the purchaser frequently obtains the bolts from one
source of supply and the nuts from another. And if they do not assemble,
which happens all too often, which supplier is at fault?

One of the more interesting features of the metric nut strength-grade system is a positive correction of this problem. It is done by establishing maxi-
mum/minimum limits for both the external and internal threads after plating
or coating and by subjecting these threads to acceptance gaging. External
threads, after coating or plating, must enter a GO thread ring gage which has

been set oversize by a specified diametral amount. The internal threads, after they have been overtapped, must be acceptable to both a GO thread plug gage, to assure ease of assembly, and a NO GO thread plug gage to demonstrate the presence of sufficient material for strength. Both plug gages are oversize by the same diametral amount as the GO thread ring gage. So, in effect, what has been done is to retain exactly the same limits of size for the internal threads of non-plated products and shift them radially outward. The relevant diametral amount was established according to the following rule:

$$A = 0.22p - 0.02$$

where

A = diametral allowance, mm
p = thread pitch, mm

Tabulated values are given in A563M. These values were carefully formulated to provide ample allowance for commercial thicknesses of galvanizing and, at the same time, to allow a convenient thread mating. Most importantly, the purchaser is always guaranteed that the bolt and nut will assemble freely regardless of source. If they do not, the offending member can easily be identified.

Only three of the metric nut property classes—5, 10S, and 12—are available overtapped. Each has its own defined strength properties and from these three there is a nut of proper strength for mating with any of the property classes of externally threaded fasteners.

A563 inch series nuts of Grades A and B do not need to be grade identification marked. Other grades such as C and C3 have three equally spaced circumferential markings on the top surface. Grades D, DH, and DH3 are identified through their grade designations. Both grades C3 and DH3 must also display the numeral "3" to indicate that they are made of atmospheric corrosion resistant steel. All of the grades, which must be grade identified, are also marked with the manufacturer's symbol to identify himself as the producer. A194 grades 2 and 2H require both grade and manufacturer identity. The corresponding grade markings are the grade designations.

SAE Grades 5 and 8 nuts are grade-identified using the, so-called, "clock" system. A dot is located at one corner on the top surface to indicate 12 o'clock and a radial line is placed at the 5 and 8 o'clock position to.identify grades 5 and 8 nuts respectively. The same marking system is, at times, recognized for grade-identifying ISO metric nuts. Nuts which are milled from bar on screw machines are normally grade-identified with small notches cut into the corners of the hex, 1 notch for Grade 5 and 2 notches for Grade 8.

All metric nuts in sizes M5 and larger are property class identified with the class designation as 5, 9, 10, 12, 8S, etc. While the marking may be located on any surface of the nut it will generally be found on the top of hex nuts and on the top of the flange of hex flange nuts. Additionally, all heat treated nuts and all structural nuts (those with "S" in their designation) must be marked to identify the manufacturer.

In Europe, and elsewhere throughout the world, the most popular nut design, prior to the finalization of the ISO Metric nut system, had a thickness of 0.8 D. Such nuts had been used successfully for many years by European industry. The reason was that their general assembly practices did not require tightening bolt/nut assemblies to the higher preload levels often used in North America. North American investigations also proved that nuts of these thicknesses, consistent with the assembly practices in North America, would have an unacceptably high risk of thread stripping.

The ISO studies initiated in the early 1970s were actually quite timely because European practice was beginning to change to more closely reflect that of North America. European fastener engineers had openly admitted the inability of their most popular nut design to meet the higher loading demands. They cooperated in research studies and were supportive of the principal conclusion that increases in nut thicknesses were absolutely necessary.

However, one giant problem remains. European standards still recognize nuts of the original design and they are being manufactured and marketed in tremendous quantities. It is still common practice to use these nuts in combination with ISO Property class 8.8 externally threaded fasteners. Unfortunately these nuts are identified with the numeral "8" as are those of the new ISO thickness.

For this reason, when the A563M standard was drafted it became the firm decision of the responsible technical committee that no class 8 nuts be recognized in the metric nut strength grade system offered to North American industry. Engineers are seriously cautioned that any metric nut identified "8" could be one of those of the old design and could have inadequate strength properties when assembled with either class 8.8 or 9.8 bolts. The best assurance of a safe design then is to make sure that none of these are used.

21

Guidelines to Bolt and Nut Selection

One of the continuously mystifying questions that seems to stump designers is which nut should be selected to mate with which bolt. Actually, the answer is relatively simple as long as we observe the following six guidelines.

1. For any low carbon steel, non-heat-treated bolt, screw, or stud (basically any fastener with a specified minimum tensile strength of 75,000 psi or less) any standard grade of *full* thickness nut is strong enough, whether it is overtapped to accomodate heavy platings or not.
2. For metric fasteners choose a nut of the same or a higher property class number (see Table 10) as the property class number of the bolt (see Table 6).
3. For inch series fasteners select a nut with a specified proof load (see Tables 8 and 9) equal to or greater than the specified minimum tensile strength of the bolt (see Tables 3 and 4).
4. For inch series fasteners, if optimum safety is a design prerequisite, select a nut with a specified proof load about 20% greater than the specified minimum tensile strength of the bolt.
5. Selection of a nut which is stronger due to additional material (i.e., greater thickness and/or width across flats) is generally less expensive

than a nut of smaller proportions which achieves its strength through heat treatment.

6. The "highest necessary" strength grade of nut offers potential cost savings.

These guidelines are examined a little more closely as follows.

Guideline 1. The lowest strength inch series full thickness nuts are A563 Grade A and SAE Grade 2 (see Tables 8 and 9) each with a proof load stress of 90,000 psi (80,000 psi for fine thread Grade A nuts). For metric application it is A563M class 5 nuts which have a proof load stress of 520 MPa or higher depending on the nut size. According to Guidelines 2 and 3 these nuts are amply strong for use with low carbon steel non-heat treated bolts (A307 and SAE Grades 1 and 2 in inch; classes 4.6, 4.8, and 5.8 in metric) with one exception. When overtapped to accomodate heavy plating thicknesses the proof load strength of these nuts drops significantly. For instance, the reduction for A563 Grade A is 25% to 68,000 psi in coarse thread nuts and down to 60,000 psi for fine thread nuts. In the case of metric A563M class 5 nuts the strength drops to 470 MPa. These overtapped nuts are still strong enough for use with galvanized inch series A307 and SAE Grade 1 bolts and for Classes 4.6 and 4.8 metric bolts but they are modestly marginal for galvanized SAE Grade 2 and class 5.8 bolts. Rarely would there be a problem in the combination, however, if the designer was to use inch series A563 Grade B and metric A563M class 12 overtapped nuts.

Guideline 2. Metric nuts were specifically designed to work properly with bolts of the same basic property class number. And, any class of metric nut is suitable for use with any class of bolt of a lower property class number. For example, class 5 nuts are adequate for use with classes 4.6, 4.8, and 5.8 bolts; class 10 nuts with class 10.9 and all lower strength bolts, etc. Conversely it is absolutely wrong to combine a metric nut with a bolt of a higher property class number; for example, a class 9 nut with a class 10.9 bolt. A study of nut proof loads (Table 11) will show that in no instance does a nut of any property class have a proof load higher than the specified minimum tensile strength of a bolt (Table 7) of a higher class number with the following exceptions. Class 8S nuts are strong enough for use with classes 9.8, and 10.9 bolts, and class 10S nuts with 12.9 bolts. But, classes 8S and 10S nuts are heavy hex structural nuts which owe much of their superior strength to their increased thickness and wider width across flats. It should be noted that these two classes of nuts are standard only in the size range M12 through M36.

Guideline 3. Over the years, as inch series nut standards evolved, nut strength grades were established by equating their proof load stresses with the specified minimum tensile strengths of the externally threaded fasteners with which they were intended to be used. Unfortunately, the dimensions of the nut and the external geometry were not always compatible. The classic case in point are SAE Grade 5 and A563 Grade B hex nuts (Tables 9 and 10) which are used in combination with SAE Grade 5 and A449 bolts. Coarse thread non-heat treated hex nuts are strong enough to meet the specified minimum tensile strengths of grade 5 and A449 bolts. Fine thread hex nuts however are not. In other words fine thread hex nuts are not always strong enough to break the bolt prior to thread stripping. For fine thread hex nuts to be as strong as grade 5 or A449 fine thread bolts, they must either be thicker or properly heat treated. Both solutions of course add expense. Actually, in most design applications SAE Grade 5 and A563 Grade B fine thread hex nuts will function satisfactorily when used with Grade 5 or A449 bolts. However, if the design calls for the bolt/nut combination to be pre-loaded to a level approaching or exceeding the bolt's specified proof load, the designer might be prudent to consider specifying SAE Grade 5 or A563 Grade B hex thick nuts (Table 8) or heat treated SAE Grade 8 or A563 Grade DH hex nuts. With this one exception any nut with a specified proof load equal to or greater than the specified minimum tensile strength of the bolt with which it is to be used should perform satisfactorily.

Guideline 4. Why is a 20% proof-load increase suggested in Guideline 4? Good design dictates that if, for any reason, the bolt/nut combination is likely to fail, either through overtightening during the assembly or over-stressing when the service loading is applied, the failure mode should be bolt fracture and not nut thread stripping. Most of the time this is exactly what will happen if the nut's proof load is equal to the specified minimum tensile strength of the bolt. In rare instances however, a nut, totally conforming to all of its requirements, but with its actual strength bordering on the permitted minimum, may be assembled with a bolt which is also completely within its specification requirements but with the strength properties close to the maximum. Both products then are perfectly acceptable against their own indivi-dual specifications but the bolt proves to be the stronger member of the team. If this combination is stressed to failure, quite conceivably the nut may strip before the bolt breaks. Consequently, if absolute assurance against this remote possibility is required the nut should have a proof load that ap-proximates the anticipated maximum tensile strength of the bolt.

Except for ASTM A490 bolts, maximum tensile strengths are not a neces-sary condition for any strength grade of externally threaded fastener. How-ever, maximum hardness is allowing a reasonable prediction of what the maximum tensile strength of the bolt might be. A study of the various strength grades will show that their estimated maximum tensile strengths are about 20% higher than their specified minima.

One of the distinct advantages of the ISO metric nut strength system is that each property class of nut was specifically designed, dimensionally and metallurgically, to properly mate with a property class of bolt. Consequently, when the correct class of nut is selected, even under the most adverse combination of conditions, the bolt will normally break first.

Guideline 5. Heat treatment is an expensive, high energy consuming process. When increased nut strength is needed and the choice is between adding material or heat treating, the former will usually be the less costly decision. For instance, consider a nut selected for a 5/8-11 SAE Grade 5 bolt with the stipulation that the nut should be 20% stronger than the bolt. This requires a nut with a proof load of 32,500 lbs, equal to strength of 1.2 X 27,100 from Table 4. Entering Table 8, the choices are an SAE Grade 8 hex nut or an A563 Grade D hex thick or heavy hex nut. The former choice refers to a heat treated conditon while the latter two are not. If space permits the accommodation of the slightly thicker hex thick nut or the thicker and wider heavy hex nut, then either of the larger nuts, will probably be less expensive than the heat treated hex nut. Of course, if there are space restraints the attention should be directed toward the heat treated, smaller nut.

Guideline 6. The use of admittedly more expensive "highest necessary" strength nuts offers an intriguing opportunity to save money in an assembly operation using several different grades of externally threaded fasteners. For example, if metric fasteners of classes 4.6, 5.8, 9.8, and 10.9 are used throughout the design and assembly of a company's products, it would be entirely practical to standardize on just class 10 nuts. In this instance class 10 is the "highest necessary" class of nut strong enough to work properly with each class of bolt being used in the company's production. The cost savings of single ordering, inventory, and handling, coupled with the complete avoidance of any assembly line bolt-nut mismatch mistakes, very generously compensates for the slightly higher original cost of using, in some combinations, a more expensive nut than actually is needed. In summary, spending additional cents for fasteners can save dollars on the production line.

22

Stainless Steel Fasteners

Stainless steels are ferrous alloys with a chromium content of at least 12%. Since chromium is an incorrodable element, all alloys within the stanless steel family have a high degree of corrosion resistance. Generally, the greater the chromium content, the more corrosion-resistant is the material.

All stainless steels, along with iron and chromium, contain carbon. And while carbon assists hardenability, unfortunately it has a deleterious effect on corrosion resistance. Carbon combines with chromium to form carbides and chromium in the form of carbides contributes negligibly to corrosion resistance. Consequently, for equivalent corrosion resistance, as carbon content increases, chromium must also be added. This is the reason why most stainless steel analyses have a very low carbon content, which is carefully controlled.

Additionally, all stainless steels have a mix of other alloying elements each of which give the alloy special characteristics and properties. Nickel is the most important element. It improves corrosion resistant properties dramatically and it adds toughness in low temperature exposures as well as strength and resistance to scaling at elevated temperatures. Other elements such as molybdenum, copper, silicon, aluminum, selenium, sulfur, tantalum, columbium, and titanium are important additives which further modify the composition of an analysis and give the product the necessary properties to meet a defined service.

The nonrusting of stainless steel is due to an ability to spontaneously form a thin, invisible oxide film completely covering its exposed surfaces. This film is passive and once formed it prevents further oxidation from occurring. During fastener manufacture, such as the heading and machining operations, the fastener's stainless steel surface may become smeared with small particles of steel abraded from the manufacturing tools and dies. If the fastener is subsequently heat treated there is also a possibility to pick up other surface contaminants. If such fasteners are then exposed in service without prior cleaning they will likely stain, not because of any corrosion of the base metal, but because of the rusting of the imbedded surface impurities. Consequently, practically all stainless steel fasteners are passivated as a final operation before shipment. Passivation is a cleaning process which consists of immersion of the fastener for several minutes in a solution of nitric acid and water. When removed from the solution the stainless steel again quickly forms an oxide film but without the entrapment of minute particles of foreign matter. For optimum corrosion resistance, stainless steel fasteners must be passivated.

Stainless steel alloys are catalogued into three general groupings: austenitic, ferritic, and martensitic. Each has its own group identity, its unique advantages, and its shortcomings. None can be viewed as the optimum, each outperforms the other depending on the particular combination of service and environmental conditions of the application, coupled with the ever important and ever present ingredient of cost.

Within each grouping there are alloy groups which have within them one or more specific analyses which are assumed reasonably interchangeable in terms of their performance characteristics. Table 12 identifies these alloy groups, for both inch and metric series fasteners, and the permissible alloys in each.

The chemical compositions, metallurgical treatments, and mechanical requirements for stainless steel fasteners are covered in our ASTM standards including various alloys discussed in this book. ASTM F593 covers inch series bolts, screws and studs with the nuts given in F594. The standards F738 and F836M cover metric, externally threaded fasteners and metric nuts. It is also important to note that the provisions of F738 and F836M are in essential agreement with ISO 3506.

The inch and metric systems are similarly structured with the various alloys catalogued into groups, metallurgical conditions, and strength grades or

Table 12 Stainless Steel Alloy Groups and Permissible Analyses

Alloy Group Designation		
For Inch Series Fasteners	For Metric Series Fasteners	Permissible Analyses
1	A1	303, 303Se, 304 305, 384, XM7[a]
2	A4	316[a]
3	A2	321, 347[a]
4	F1	430, 430F[b]
5	C1	410[c]
	C4	416, 416Se[c]
6	C3	431[c]
7	P1	630[d]

[a]Types 303, 303Se, 304, 305, 384, XM7, 316, 321, and 347 are austenitic steels
[b]Types 430 and 430F are ferritic.
[c]Types 410, 416, 416Se and 431 are martensitic steels.
[d]Type 630 is a precipitation hardening steel.

property classes. Metric property classes are identified using a fully significant designation system. The inch series designation system however is not as definitive.

For externally threaded fasteners the principal mechanical properties are tensile strength, yield strength, elongation, and hardness. Proof loads are not a specification requirement. The reason for this is that, unlike in the carbon steels, the stress-strain behavior of stainless steel is not as predictable nor has it a "straight line" consistency. A typical stainless steel exhibits a "round house" stress-strain curve with a small degree of permanent deformation occurring at a relatively low tensile stress. Consequently, yield strength, the stress at which an offset equal to 0.2% of the gage length occurs in the stress-strain curve is the basic mechanical requirement.

Proof loads and hardnesses are the required properties for nut selection. Inch series hex nuts (ANSI B18.2.2) and metric hex style 1 nuts (ANSI B18.2.4.1M) are dimensionally adequate for all strength grades and property classes. It is normal practice to combine bolts and nuts of the same alloy and condition for inch series and of the same property class for the metric specification.

Inch series fasteners need only be grade and manufacturer identification marked when the customer requests. The metric standards, including ISO 3506, require that hex head bolts and screws and all nuts, in sizes M5 and larger, be marked with the property class designation and the manufacturer's symbol.

And now one last comment about stainless steel fasteners. Stainless steel is not impervious to stress embrittlement. While not as susceptible as carbon steels, failures can and do occur. Rarely will there be a problem with any of the austenitic or ferritic steel fasteners but sometimes the higher strength martensitic alloys may experience embrittlement depending on the severity of the exposure and the magnitude of the induced stresses due to tightening or application of the service loads. If the designer has any suspicion of a potential problem he is well advised to consult with a metallurgist experienced in the behavior of stainless steel.

23

Advantages and Limitations of Austenitic Stainless Steels

Austenitic stainless steels are commercially nicknamed the 18-8 series or the 300 series. They are characterized by having chemical compositions containing about 18% chromium and 8% nickel. These steels as a group exhibit better corrosion resistance than either the ferritic or martensitic steels and are nonmagnetic. These materials do not respond to heat treatment, although their mechanical properties can be significantly improved through cold working. They have excellent strength properties and toughness at extremely low temperatures and are known to perform very respectably in elevated temperatures.

About 80% of all stainless steel fasteners are produced from one of the austenitic grades. While there are 25 or more standard analyses and as many or more proprietaries, only nine have any degree of popularity in the manufacture of fasteners. And these fall into three groupings with the alloys in each group having similar performance properties considered to be generally interchangeable. In the first group there are Types 305, XM7, 303, 303Se, 304, and 384. The second group is reserved for Type 316 while Types 321 and 347 constitute the third grouping.

For many years Types 305 (17/19% Cr, 10.5/13% Ni) has been the workhorse because its higher nickel content slows the rate of work hardening

Table 13 Mechanical Properties of Stainless Steel for Inch Series Fasteners

Alloy Group	Condition[a]	Nominal Size of Fastener (in.)	Full Size Bolts Screws and Studs Tensile Strength (ksi) Min	Yield Strength[d] (ksi) Min	Machined Test Specimens[b] Tensile Strength (ksi) Min	Yield Strength[d] (ksi) Min	Elongation (%) Min	Nut Proof Load Stress[c] (ksi)	Hardness Rockwell Min	Max
1,2,3	A	1/4 to 1 1/2	75	30	70	30	30	75	B65	B95
	AF	1/4 to 1 1/2	85 max	–	80 max	50 max	40	70	–	B85
	CW	1/4 to 5/8	100	65	95	60	20	100	B95	C32
		3/4 to 1 1/2	85	45	80	40	25	85	B80	C32
	SH	1/4 to 5/8	120	95	115	90	12	120	C24	C36
		3/4 to 1	110	75	105	70	15	110	C20	C32
		1 1/8 to 1 1/4	100	60	95	55	20	100	B95	C30
		1 3/8 and 1½	95	45	90	40	28	95	B90	C28
4	A, AF	1/4 to 1 1/2	70	35	70	35	25	70	B65	B95
5	H	1/4 to 1 1/2	110	90	110	90	18	110	C20	C30
	HT	1/4 to 1 1/2	160	120	160	120	12	160	C34	C45

6	H	1/4 to 1 1/2	125	100	125	100	15	125		C25	C32
	HT		180	140	180	140	10	180		C40	C48
7	AH	1/4 to 1 1/2	135	105	135	105	16	135		C28	C38

[a]Legend of conditions:

A, the fastener is machined from annealed stock thus it retains the properties of the raw material

AF, the fastener is fully annealed after its manufacture

CW, the fastener is cold made and allowed to work harden; fasteners in sizes larger than 5/8 in. may be hot worked

SH, the fastener is severely cold worked or is machined from strain hardened stock

H, the fastener is hardened and tempered at 1050°F (565°C) minimum

HT, the fastener is hardened and tempered at 525°F (275°C) minimum

AH, the fastener is solution annealed and age hardened

[b]Externally threaded products with specified tensile strengths of 100,000 lbs and less are tested full size. Products with higher specified tensile strengths are preferably tested full size but may be tested using machined test specimens.

[c]Nuts with proof loads of 120,000 lbs and less are proof load tested. Nuts with higher proof loads are normally accepted on the basis of meeting their hardness requirements.

[d]Yield strength is stress at which an offset of 0.2% of gage length occurs.

[e]To compute tensile strengths, yield strengths and nut proof loads in pounds, multiply the applicable stress by the thread stress area as given in Table 1.

Table 14 Mechanical Properties of Stainless Steel Metric Fasteners

Property Class	Condition[a]	Nominal Thread Diameter	Full Size Bolts Screws and Studs		Machined Test Specimens[b]			Nut Proof Load Stress[c]	Hardness Rockwell	
			Tensile Strength (MPa) Min	Yield Strength[d] (MPa) Min	Tensile Strength (MPa) Min	Yield Strength[d] (MPa) Min	Elongation (%) Min	MPa	Min	Max
A1-50 A2-50 A4-50	A or AF	M6 to M36	500	210	500	210	30	500	B81	B95
A1-70 A2-70 A4-70	CW	M6 to M20 over M20 to M36	700 550	450 300	650 520	400 270	20 25	700 550	B96 B83	C33 C31
A1-80 A2-80 A4-80	SH	M6 to M20 over M20-M24 over M24-M30 over M30-M36	800 700 650 600	600 500 400 300	780 680 620 570	600 480 370 270	12 15 20 28	800 700 650 600	C23 B96 B93 B89	C36 C33 C30 C28

Property class	Grade	Thread diameter					Elong., %		Hardness	
F1-45	AF	M6 to M36	450	250	450	250	25	450	B74	B96
C1-70 C4-70	H	M6 to M36	700	410	700	410	18	700	B96	C34
C1-110 C4-100	HT	M6 to M36	1100	820	1100	820	12	1100	C36	C45
C3-80	H	M6 to M36	800	640	800	640	15	800	C23	C35
C3-120	HT	M6 to M36	1200	950	1200	950	10	1200	C39	C48
P1-90	AH	M6 to M36	900	700	900	700	16	900	C28	C38

aSee footnote 1 to Table 13.

bExternally threaded products with specified tensile strengths of 450 kN and less are tested full size. Products with higher specified tensile strengths are preferably tested full size but may be tested using machined test specimens.

cNuts with proof loads of 530 kN and less are proof load tested. Nuts with higher proof loads are normally accepted on the basis of meeting their hardness requirements.

dYield strength is stress at which an offset of 0.2% of gage length occurs.

eFor mechanical properties of screws and nuts of sizes smaller than M6 refer to ASTM F738 and Fxxx respectively.

fTo compute tensile strengths, yield strengths and nut proof loads in kilonewtons, multiply the applicable stress by the thread stress area as given in Table 2 and divide by 1000.

which in turn makes cold heading and forming easier. In recent years Type XM7 (17/19% Cr; 8/10% Ni; 3/4% Cu) has become very popular. XM7 is basically Type 302 modified to improve its cold headability. It is also an outgrowth of Armco's proprietary steel 18-9LW. The reasons for the growing popularity of this steel include superior cold formability and the reported lower cost than that of 305 or 384. Types 303, and 303Se (17/19% Cr; 8/10% Ni) are free-machining grades which have fairly good hot forging properties but are not particularly suitable for cold heading. These materials are frequently chosen for large size nuts which require extensive machining operations. Type 304 (18/20% Cr; 8/10.5% Ni; 0.08% C max) is a low carbon grade. The low-carbon, high-chromium ratio improves its corrosion resistant characteristics. It has good hot and cold heading properties and it is frequently used for larger size fasteners which must be hot made.

Type 384 (15/17% Cr; 17/19% Ni; 0.08% C max) is an outgrowth of Carpenter Alloy No. 10. It is an austenitic stainless steel developed specially for severe cold heading upset applications such as cross recessed screws and cold forged nuts. The exceptionally high nickel content impedes the rate of work hardening but it adds considerably to the raw material cost. Type 384 has stable, nonmagnetic properties following cold heading. The majority of other austenitic steels however become slightly magnetic when cold worked and must be annealed to restore their non-magnetic characteristics.

Type 316 (16/18% Cr; 10/14% Ni; 2/3% Mo; 0.08% C max) has significantly improved corrosion resistance due to the inclusion of molybdenum which gives the steel a quality to resist surface pitting. This alloy also has higher tensile and creep strengths at elevated temperatures than those of most other austenitic steels. It readily upsets both cold and hot although its machining qualities are inferior. Finally the higher nickel and molybdenum content makes Type 316 a bit more expensive.

Types 321 (17/19% Cr; 9/12% Ni) and 347 (17/19% Cr; 9/13% Ni) are stabilized austenitic steels. Type 321 is stablized using titanium while type 347 requires columbium and tantalum for this process. The "stabilizing" eliminates carbide precipitation and consequently intergranular corrosion. This gives these steels excellent corrosion resistant properties at elevated temperatures of up to about 1500°F. They are particularly suited for aerospace fasteners and those used in high temperature chemical processing environment.

It should be pointed out that austenitic stainless steel fasteners cannot be heat treated. Even so, exceptionally high strength properties can be achieved through cold working. Fasteners, both externally and internally threaded, are generally available in the following four conditions:

AF: The fastener is fully annealed following its manufacture.

A: The fastener is machined from annealed stock with no further treatment after its manufacture. Hence the fastener in this condition retains the properties of the raw material.

CW: The fastener is cold headed and roll threaded and allowed to work harden, fasteners in sizes larger than 3/4 in., (20 mm) are usually hot headed, which reduces their strength properties to approximately those of the annealed fasteners.

SH: The fastener is severely cold worked or machined from strain hardened stock.

Austenitic stainless steel fasteners are normally supplied in the cold worked (CW) condition unless the customer specifically orders otherwise. Annealed fasteners (AF or A) may be specified if the prime design criterion is optimum corrosion resistance and not strength. Strain hardened (SH) fasteners are specified when the basic strength is important. In fact, strengths of strain-hardened fasteners in the smaller sizes compare quite respectably with those of carbon steel Grade 5 and ASTM A449. Some manufacturers, employing advanced cold working techniques, can successfully cold head fasteners in sizes as large as 5/8 in. (16 mm) and achieve the strain hardened properties. Larger sizes usually have to be machined from strain hardened bars and this adds tremendously to their cost. In practice seldom does the cost justify the relatively modest strength differential existing between large size CW and SH fasteners. The strength properties of austenitic stainless steel for inch series fasteners are detailed in Table 13. The properties of metric fasteners are compiled in Table 14.

24

Use of Ferritic Stainless Steels

Ferritic stainless steels are iron-chromium alloys in which the presence of any nickel is purely incidental. They do not respond to heat treatment, nor do they exhibit strength appreciation through cold working. In fact, fasteners of these alloys are normally furnished fully annealed because while cold working increases strengths and hardnesses, such increases are accompanied by disproportionately large sacrifices in ductility. Although ferritic steels have superior corrosion resistance to martensitic steels in most environments they are not as versatile in their corrosion resistant properties as austenitic steels. Furthermore ferritic stainless steels are fully magnetic in all conditions. They have reasonably good resistance to scaling at moderately high temperatures and in comparison with austenitic steels their coefficient of thermal expansion is significantly lower.

Ferritic stainless steels account for about 5% of all stainless steel fasteners produced and practically all are made from Type 430 (14/18% Cr; 0.12% C max). This alloy has reasonably good cold heading and hot forging properties. However if the parts require extensive machining then Type 430F, with its "machinability assisting" sulfur content, is a better choice. Economy, low cost of raw material, coupled with good corrosion resistant properties, are the principal recommendations for the selection of ferritic stainless steel for fasteners. Their strength properties are detailed in Tables 13 and 14.

Use of Ferritic Stainless Steels

25

Use of Martensitic and Precipitation Hardening Stainless Steels

The main characteristic of martensitic stainless steel is its ability to respond to heat treatment in a similar fashion to that of a carbon steel. Consequently, martensitic stainless steel fasteners are high strength products with the properties approximating those of SAE grades 5 and 8, ASTM Grade A449, and A354 grade BD.

Martensitic stainless steels are iron-chromium alloys with chromium content in the range of 12 to 17% as well as sufficient carbon content to assure heat-treatability. While their corrosion resistance may not be quite as good or as versatile as that for either the austenitic or ferritic steels, their performance in the heat treated condition is quite adequate for most atmospheric and marine exposures. Strength properties at temperatures up to about 1100°F are also dependable, although, unlike austenitic steels, they lose their toughness when exposed to subzero climates. Furthermore martensitic stainless steels are magnetic in all conditions. Despite some limitations the fasteners made of these alloys offer an attractive combination of strength, corrosion resistance, and economy.

Perhaps as many as 10% of stainless steel fasteners are produced using martensitic steels with Types 410, 416, 416Se, and 431 being selected almost exclusively. Type 410 (22.5/13.5% Cr; 0.15% C max) has strength

115

properties roughly equivalent to those of SAE Grade 5 or A 449 although, through special treatments, its properties can be higher. Type 410 cold-heads and hot-forges well, and, because of its low chromium content it is the least expensive of all the stainless steels. It should be considered a prime candidate for use in any application where zinc or cadmium plated Grade 5 fasteners are insufficient to provide the necessary resistance to corrosive attack.

Types 416 and 416Se (12/14% Cr; 0.15% C max; with sulfur and/or selenium) are possibly the most readily machinable of all stainless steels. Their mechanical properties parallel those of Type 410. However the addition of sulfur and selenium to assist machinability modestly disadvantages their corrosion resistant properties. At the same time the presence of these elements helps to reduce any tendency to seize or gall during the fastener tightening or in service.

Type 431 (15/17% Cr; 1.25/2.5% Ni; 0.20% C max) is used for fasteners with strengths approaching, and when specially processed, exceeding those of carbon alloy steel SAE Grade 8 and ASTM A354 Grade BD. In fact, tensile strengths greater than 200 ksi are entirely practical. Type 431 is frequently selected for aerospace fasteners. It has the best corrosion resistant properties of the martensitic steels and it readily cold-forms and hot-heads. However it is rather difficult to machine.

The strength properties of fasteners made of these martensitic stainless steel alloys are compiled in Tables 13 and 14.

Close to 5% of all stainless steel fasteners are made of precipitation hardening steels and their popularity is growing rapidly. These steels are characterized by having the excellent corrosion resistance of austenitic steels and the high strength capabilities of the martensitic steels. Type 630 (15.5/17.5% Cr; 3/5% Ni; 0.07% C max; 0.15/0.45% Cb plus Ta), also known as 17-4 PH, is probably the precipitation hardening steel most extensively used in the manufacture of fasteners. It has superior ductility and its strength properties, which parallel those of heat treated Type 410, are achieved through solution annealing and controlled age hardening following fastener manufacture. The relevant service performance in both high and low temperatures is reasonably good. The mechanical properties of Type 630 fasteners are given in Tables 13 and 14.

26

Applications of Copper Base Alloys

Over 220 different copper base alloys are commercially available for the designer's selection and use in various engineering applications. Of these, only about six enjoy any significant degree of popularity as a material for mechanical fasteners, with perhaps 10 others being considered as infrequent alternatives.

Copper offers a number of interesting performance features which singly or in combination may identify it, or one of its alloys, as exactly the right material for a specific application. Its thermal and electrical conductivity are the best of any of the nonprecious metals. The corrosion resistance of copper in most environments is quite respectable and it is totally nonmagnetic. Many copper alloys have reasonably high strength, are malleable, and offer a wide range of attractive colors with the ability to accept and retain a high surface luster. Among the shortcomings are relatively low strength-to-weight ratio, severe loss of strength in low temperatures, and an unfortunately high susceptibility to stress corrosion cracking. Many of the copper alloys gain strength through cold working and age hardening. However, in order to obviate the risk of stress corrosion it is frequently necessary to stress relieve the fastener after its manufacture. This process in turn nullifies the potential strength improvements.

117

Pure copper is defined as having a minimum copper content of 99.3%. Copper alloys are characterized by at least 40% copper content. Brasses are copper alloys in which zinc is the main alloying element. Copper-nickel alloys have nickel as their principal alloying element. And bronzes are those copper alloys in which neither zinc nor nickel are the dominant alloying elements.

Pure copper is infrequently used as a fastener material because of its low strength and higher cost. However, in its pure form copper exhibits its optimum conductivity properties and as alloying elements are added, conductivity generally declines. Alloy No. 110, Electrolytic Tough Pitch Copper (99.9% Cu), is a typical composition. It is highly malleable and it forms very well in cold and hot conditions. Pure copper is ideally suited for small rivets, washers, and other "nonstructural" fasteners.

The brasses are perhaps the most popular family of copper alloys. They retain most of the favorable characteristics of copper and add some new ones at a lower cost. The magnitude of copper content is important. Small percentage differences can cause interesting, and sometimes gross, changes in the material's character. Alloys with less copper content are usually stronger and harder but less ductile. As copper is added, headability, particularly cold, improves. Naval brass, Alloy No. 462 (63.5% Cu; remainder Zn) with its excellent cold-heading characteristics and Alloy No. 464 (60% Cu; remainder Zn) with its equally good hot-forming properties are extremely popular. Yellow brass, Alloy No. 270 (65% Cu; remainder Zn) is used extensively in the manufacture of milled-from-bar nuts. Free machining brass, Alloy No. 360 (61.5% Cu; 3% Pb; remainder Zn) is by far the most popular for small screw machine parts. Manganese bronze, Alloy No. 675 (58.5% Cu; 1.4% Fe; 1% Sn; remainder Zn) could be classified as a brass because of its high zinc content. The modest amounts of tin, iron, and manganese help its strength properties without the sacrifice of either thermal or electrical conductivity.

The silicon bronzes are the most popular of the bronzes with Alloy No. 651 (98.5% Cu; 1.5% Si) being the best choice for cold headed parts. Alloy No. 655 (97% Cu; 3% Si) is the appropriate material for fasteners which must be hot formed. Alloy No. 661 (95% Cu; 3% Si; 0.5% Pb) is not too popular but is sometimes chosen because of its better machining qualities. The silicon bronzes have good strength and toughness which, coupled with their corrosion resistance and nonmagnetic properties, ideally suit these alloys for general naval construction and minesweepers in particular. Cold formed silicon bronze fasteners should be stress relieved to reduce the risk of stress corrosion failure.

Aluminum bronzes are a family of copper alloys with 4 to 10% of aluminum in their composition plus small amounts of other alloying elements added to improve tensile strength, toughness and ductility. Alloy No. 613 (92.7% Cu; 7% Al; 0.3% Sn) and Alloy No. 614 (91% Cu; 7% Al; 2% Fe) are ideally suited for cold heading with perhaps 613 being the better choice. Alloy No. 630 (82% Cu; 10% Al; 5% Ni; 3% Fe) is frequently chosen when parts are of a size that necessitates they be made hot; and Alloy No. 642 (91.2% Cu; 7% Al; 1.8% Si) is an excellent material for hot made fasteners which require secondary machining operations. While the tensile strength of 630 externally threaded fasteners appears quite high compared to other copper alloys its ductility is very low. Aluminum content greatly assists resistance to scaling and oxidation at elevated temperatures and all of the aluminum bronzes exhibit good corrosion resistance in turbulent sea water, which is considered to be an extremely aggressive environment. Their corrosion resistance in contact with mineral acids is generally quite good although only fair performance is expected in contact with strong alkalies.

Phosphor bronze, Alloy No. 510 (94.8% Cu; 0.2% P; 5% Sn), has strength, toughness and good fatigue resistant properties. The small addition of phosphorus serves as a deoxidizing agent which helps against stress corrosion while tin is a hardening agent which boosts the mechanical strength. No. 510 has excellent cold working abilities. However its hot formability and machining properties are well below average.

The distinctive characteristic of copper-nickels is their superior resistance to corrosion and erosion in high velocity sea water. Their high nickel content not only contributes significantly to corrosion protection but enhances toughness, ductility and resistance to stress corrosion and corrosion fatigue. However, it adds cost. Alloy No. 710 (79% Cu; 21% Ni) and Alloy No. 715 (69.5% Cu; 30% Ni; 0.5% Fe) are the materials frequently used in fastener manufacture. Their cold and hot fabrication properties are reasonably good although their ability to be machined is poor. While copper-nickels have interesting capabilities, their cost is high in comparison with other copper alloys and stainless steels.

The copper alloy number designation system was originally developed by the United States copper and brass industry and it is administered by its Copper Development Association (CDA). Several years ago, ASTM and SAE jointly designed and currently manage the Unified Numbering System (UNS) for Metals and Alloys. UNS is a comprehensive and logical method of assigning identification numbers to the alloys of all base metals. All of the copper alloys are covered within the UNS system where the appropriate UNS num-

Table 15 Mechanical Properties of Copper Alloy Fasteners for Inch and Metric Series[a]

Copper Alloy Number	General Name	Full Size-Externally Threaded Products[e]				Machined Test Specimens					Nut Proof Load Stress		Hardness Rockwell	
		Tensile Strength Min		Yield Strength[c] Min		Tensile Strength Min		Yield Strength Min		Elongation[d] % Min			Min	Max
		ksi	MPa	ksi	MPa	ksi	MPa	ksi	MPa		ksi	MPa		
110	ETP Copper	30	205	10	70	30	205	10	70	15	30	205	F65	F90
270	Yellow Brass, 65%	60	410	50	345	55	380	50	345	35	60	410	F55	F80
360	Free Cutting Brass	55	380	40	275	55	380	40	275	25	55	380	B60	B90
462	Naval Brass	50	345	25	170	50	345	25	170	20	50	345	B65	B90
464	Naval Brass	50	345	15	105	50	345	15	105	25	50	345	B55	B75
510	Phosphor Bronze, 5%A	60	410	35	240	55	380	30	205	15	60	410	B60	B95
613[b]	Aluminum Bronze	80	550	50	345	80	550	50	345	30	80	550	B70	B95
614	Aluminum Bronze D	75	520	35	240	75	520	35	240	30	75	520	B70	B95

630	Aluminum Bronze	100	690	50	345	100	690	50	345	5	100	B85	B100
642	Aluminum Silicon	75	520	35	240	75	520	35	240	10	75	B75	B95
651[b]	Low Silicon Bronze B	70	480	55	380	70	480	53	365	8	70	B75	B95
655	High Silicon Bronze A	50	345	20	140	50	345	15	105	20	50	B60	B80
661	Silicon Bronze	70	480	35	240	70	480	35	240	15	70	B75	B95
675	Manganese Bronze A	55	380	25	170	55	380	25	170	20	55	B60	B90
710	Copper Nickel, 20%	45	310	15	105	45	310	15	105	40	45	B50	B85
715	Copper Nickel, 30%	55	380	20	140	55	380	20	140	45	55	B60	B95

aThe tabulated values apply to inch series products in nominal sizes 1/4 through 1-1/2 in. and to metric products with nominal thread diameters M6 through M36.

bThe tabulated stress values apply only to sizes 3/4 in., M20, and smaller. For larger sizes, which normally must be made hot, stresses are approximately 25% lower.

cYield strength is the stress at which an offset of 0.2% of gage length occurs.

dElongation is based on a gage length equal to 4 times the diameter of the test specimen.

eYield and tensile strengths (in pounds and kilonewtons) of externally threaded fasteners and proof loads of nuts may be computed by multiplying the applicable stress by the tensile stress area of the thread as given in Tables 1 and 2.

bers are simply extensions of their previously assigned CDA numbers by pre-
fixing with the letter "C" and suffixing with "00". For example, CDA Alloy
No. 675 became C67500 in the UNS system. UNS numbers are used extensively throughout technical standards and literature. However it will be several years before the UNS system becomes the accepted norm in commercial
parlance with the more familiar numbers and designations being phased out.

Four ASTM standards cover the strength properties of copper and copper
alloy fasteners. ASTM F467 specifies requirements for inch series nuts, and
F468 those for inch series bolts, screws, and studs. Their metric counterparts, both of which are "soft" conversions, are F467M and F468M, respectively. There are no ISO standards defining the properties of copper alloy
fasteners, nor are any contemplated.

Table 15 presents the strength properties of internally and externally
threaded fasteners, both inch and metric series, and for all of the alloys mentioned in the preceding discussion. These values agree with those given in the
ASTM standards. It should be noted however that the chemical compositions
included with each alloy in the text are merely nominals. Complete analyses,
with toleranced limits for each element, are given in the ASTM standards and
in the technical literature of the CDA.

Copper alloy fasteners offer the engineer some rather remarkable combinations of performance attributes although the selection of the optimum alloy
for a specific application is not always an easy or obvious design decision.
When in doubt, consult an experienced metallurgist, or a manufacturer
specializing in the production of nonferrous fasteners.

27

Special Features of Aluminum Alloys

Of all the many basic nonferrous materials selected for engineering applications, aluminum is the most popular. Aluminum is synonymous with light weight and its use is destined to grow with expanding engineering emphasis on energy conservation through weight reduction. Numerous aluminum components can also be most properly assembled using aluminum fasteners.

Aluminum fasteners weigh about one-third those os steel with the strength properties and the strength-to-weight ratios of the most frequently selected alloys being much better than any other commercially used fastener material. Additionally, aluminum has the following attractive features. It is nonmagnetic, and possesses good thermal and electrical conductivities equal to about two-thrids those of copper. Furthermore, aluminum is machineable and it can easily be cold formed and hot forged. Aluminum's corrosion resistant properties are adequate in most ordinary environments and can be significantly improved through anodizing for the particularly aggressive exposure. Anodizing is an electrolytic process which builds an oxide film on the surface which not only adds corrosion protection but enhances resistance against wear and abrasion. Anodic coatings can be of various colors for decorative or identification purposes. In a corrosion-inducing atmosphere aluminum forms a light grey surface coating of aluminum oxide although unlike many other metals under corrosive attack, the corrosion products never stain or creep onto the adjacent surfaces.

Aluminum alloys, of which there are about 100, are designated using a four digit numerical system administered by the Aluminum Association. The system is significant in that the first digit of the four identifies a basic group of alloys. The last two digits designate the separate alloys within the group while the second digit is reserved for alloy modifications. For example, alloys in the 1xxx group are essentially pure aluminum, those designated 2xxx have copper as their principal alloying element, 5xxx indicate magnesium, 6xxx denote magnesium and silicon, while 7xxx designation refers to aluminum alloyed with zinc.

Suffixed to the alloy number is a letter/numeral(s) designation which identifies temper or a metallurgical treatment. A letter denotes the basic temper while a numeral(s) refers to the specific sequence of treatments to arrive at that temper. For example, "0" means fully annealed, "F" means as fabricated, and "T" indicates heat treated. Heat treatment tempers are "T3" which is solution heat treated and then cold worked, "T4" which is solution heat treated and naturally aged to a substantially stable condition, "T6" which is solution heat treated and then artificially aged, "T9" which is solution heat treated, artificially aged and then cold worked, and "T73" which is a proprietary heat treatment developed by Alcoa.

In the Unified Numbering System for Metals and Alloys the UNS designations for aluminum alloys are those originally assigned by the Aluminum Association with each alloy prefixed with "A9". For example, alloy 2024 became UNS Alloy A92024.

Pure aluminum has a tensile strength of about 13,000 psi. However by adding small amounts of alloying elements tremendous increases in strengths are possible. Alloys in the 2xxx, 6xxx, and 7xxx groups respond well to heat treatment and consequently practically all threaded fasteners intended for use in load transmitting applications are made of alloys from these three groups. In fact, just four alloys are used almost exclusively.

Alloy 2024-T4 (4.5% Cu; 0.6% Mn; 1.5% Mg; remainder Al) is the workhorse. Its well balanced combination of strength, corrosion resistance, fabrication properties, and economy make it the heavy favorite for externally threaded fasteners and some nuts. Bolts, screws, and studs of alloy 7075-T73 (1.6% Cu; 2.5% Mg; 0.3% Cr; 5.6% Zn; remainder Al) are slightly stronger and when subjected to the special T73 treatment they become largely impervious to stress corrosion cracking. Their popularity however is somewhat limited because of the modestly higher cost. Alloy 6061-T6 (0.6% Si; 0.25%

Cu; 1% Mg; 0.2% Cr; remainder Al) is used for both externally and internally threaded fasteners when greater corrosion resistance is a design prerequisite. Alloy 6262-T9 (0.6% Si; 0.25% Cu; 1% Mg; 0.09% Cr; 0.5% Pb; 0.5% Bi; remainder Al) is recommended almost exclusively for nuts. This alloy is stronger than 6061-T6 and has comparable corrosion resistant properties. Full thickness nuts made of 6262-T9 are generally of adequate strength and can be used with the bolts fabricated from either 2024-T4 or 7075-T73. Machine screw nuts and other styles of small nuts of sizes 1/4 in. (M6) and smaller are usually made out of 2024-T4.

The strength properties of inch-series aluminum alloy fasteners are covered in ASTM standards F467 for nuts and F468 for externally threaded products. Properties of metric nuts are included in ASTM F467M and those of metric bolts, screws and studs in F468M. Table 16 presents the basic strength data for.the inch and metric series of fasteners.

Two points of departure between the mechanical properties of aluminum alloy externally-threaded fasteners and those made from other metallic materials are worth mentioning. The first is that the load carrying capability of the part is predicated on the cross-sectional area at the root of its thread and not the larger tensile stress area corresponding to the cross-sectional area at the thread pitch diameter. In Table 16 the tensile and yield strengths in pounds or kilonewtons specified for the machined test specimens are "true" strength values for the particular alloy. Those specified for full size fasteners have been adjusted so that when the stresses are multiplied by the tensile stress area the computed values are approximately the same as those obtained by multiplying the "true" stresses by the smaller thread root area.

The second difference is that hardness of aluminum has very little significance and is generally meaningless as a criterion for inspection and acceptability. As a substitute (and this technique is frequently used to mechanically test fasteners which are too short to be tensile tested) the fastener is shear tested. For this reason shear values are included in Table 16 without any reference to hardness.

While the foregoing alloys are most suitable for load carrying fasteners, many other aluminum alloys can be used in fastener manufacture. For instance small solid, semi-tubular, and blind rivets are often made of 1100-F, 5052-F, 5056-F, the heat treatable alloys 2017-T4, 2117-T4, 2024-T4, 6061-T6, and the relatively new alloy 7050-T73, which has superior shear strength and can still be driven as received without need for predriving treatment. Flat washers are made of alclad 2024-T4 while helical spring washers are fabricated in 7075-T6. Aluminum tapping screws are available made of

Table 16 Mechanical Properties of Aluminum Alloy Fasteners for Inch and Metric Series[a]

| Alloy Number and Temper | Full Size Externally Threaded Products[a] | | | | Machined Test Specimens | | | | | | | | | Nut Proof Load Stress | |
| | Tensile Strength[d] Min | | Yield Strength[b,d] Min | | Tensile Strength[b] Min | | Yield Strength[b] Min | | Elongation[c] (%) Min | Shear Strength[e] | | | | | |
	ksi	MPa	ksi	MPa	ksi	MPa	ksi	MPa		ksi	MPa			ksi	MPa
2024-T4	55	380	36	250	62	430	40	275	10	37	255			55	380
6061-T6	37	255	31	215	42	290	35	240	10	25	170			40	275
6262-T9	–	–	–	–	–	–	–	–	–	–	–			52	360
7075-T73	61	420	50	345	68	470	56	385	10	41	280			–	–

[a]The tabulated values apply to inch series products in nominal sizes 1/4 through 1-1/2 in. and to metric products with nominal thread diameters M6 through M36.

[b]Yield strength is the stress at which an offset of 0.2% of gage length occurs.

[c]Elongation is based on a gage length equal to 4 times the diameter of the test specimen.

[d]Yield and tensile strengths (in pounds and kilonewtons) of externally threaded fasteners and proof loads of nuts may be computed by multiplying the applicable stress by the tensile stress area of the thread as given in Tables 1 and 2.

[e]Shear load (in pounds and kilonewtons) of externally threaded fasteners may be computed by multiplying the shear stress by the cross-sectional area of the shear plane. When the shear plane occurs through the threaded length the cross-sectional area is assumed to be the thread root area as given in Tables 1 and 2.

7075-T6 and self-drilling screws of the same alloy but anodized. The anodizing is applied in order to harden the surface for the drilling and tapping operations. The most popular alloy for screw machine parts is 2011-T3 (5.5% Cu; 0.5% Pb; 0.5% Bi; remainder Al).

28

Nickel Based Alloys

The family of nickel alloys offer some truly remarkable combinations of performance capabilities. Mechanically these alloys have good strength properties, exceptional toughness and ductility, and are generally immune to stress corrosion. Their corrosion resistant properties and performance characteristics in both elevated and subzero temperatures are superior. Also they are nontoxic. Unfortunately, nickel based alloys are relatively expensive.

By far the two most popular nickel alloys used for fasteners are of the nickel-copper and nickel-copper-aluminum type. Nickel-copper alloy (67% Ni; 30% Cu; 1.4% Fe; 1% Mn), known commercially by such trade names as Monel, Harper 400, Cunel, and D-H 400, optimally combines strength, toughness, corrosion resistance, and economy. Although this alloy is non-heat-treatable its strength properties can be improved through cold working. Nickel-copper class A is particularly suitable for cold heading. Class B alloy because of its slightly higher sulfur content, is a free machining grade. Small size fasteners made of class A are modestly stronger because of the cold work. Otherwise, fasteners of class A and class B have essentially the same chemical composition and exhibit similar performance characteristics.

Nickel-copper-aluminum alloy (66% Ni; 29% Cu; 1% Fe; 0.75% Mn; 2.75% Al; 0.6% Ti), commercially tradenamed K-Monel, is an extension of a nickel-

Table 17 Mechanical Properties of Nickel Alloy Fasteners for Inch and Metric Series

Nickel Alloy Number	General Name	Nominal Size of Product (in./mm)	Full Size-Externally Threaded Products				Machined Test Specimens				Elong-ation[c] (%) Min	Nut Proof[d] Load Stress		Hardness Rockwell	
			Tensile[d] Strength Min		Yield[d] Strength[b] Min		Tensile Strength Min		Yield Strength[b] Min						
			ksi	MPa	ksi	MPa	ksi	MPa	ksi	MPa	Min	ksi	MPa	Min	Max
400	Nickel-Copper Class A	1/4 (M6) thru 3/4 (M20)	80	550	40	275	80	550	40	275	20	80	550	B75	C25
		7/8 (M24) thru 1 1/2 (M36)	70	480	30	205	70	480	30	205	20	80	550	B60	C25
400HF[a]	Nickel-Copper Class A	1/4 (M6) thru 1 1/2 (M36)	70	480	30	205	70	480	30	205	20	80	550	B60	B95
405	Nickel-Copper Class B	1/4 (M6) thru 1 1/2 (M36)	70	480	30	205	70	480	30	205	20	70	480	B60	C20
500	Nickel-Copper -Aluminum	1/4 (M6) thru 3/4 (M20)	130	900	90	620	130	900	90	620	20	130	900	C24	C37
		7/8 (M24) thru 1 1/2 (M36)	130	900	85	590	130	900	85	590	20	130	900	C24	C37

[a] The properties specified for Alloy 400 HF apply only to products which are hot formed.
[b] Yield strength is the stress at which an offset of 0.2% of gage length occurs.
[c] Elongation is based on a gage length equal to 4 times the diameter of the test specimen.
[d] Yield and tensile strengths (in pounds and kilonewtons) of externally threaded fasteners and proof loads of nuts may be computed by multiplying the applicable stress by the tensile stress area of the thread as given in Tables 1 and 2.

copper alloy. The aluminum and titanium elements improve the response to heat treatment and significantly enhance the mechanical strength. In fact the strength properties are close to those of carbon steel in SAE Grade 8.

Nickel-copper is magnetic at room temperature while K-Monel is not. Both alloys have exceptional subzero properties with their strengths actually increasing slightly with lowering the temperature but without any appreciable effect on ductility, hardness, or toughness. The UNS designations of nickel-copper classes A and B, and nickel-copper-aluminum alloys are N04400, N04405 and N05500, respectively.

The mechanical properties of nickel alloy for inch and metric series fasteners are contained in ASTM standards. F467 covers the inch-series nuts, F468 inch-series externally threaded products while the F467M and F468M specifications relate to their metric counterparts. Table 17 is a compilation of mechanical properties for inch and metric series of fasteners fabricated from nickel alloys.

29

Titanium Alloys

Titanium, as an engineering material, is basically a development of the last 40 years. While this may seem a long time, actually, in materials engineering, it is a relatively short period in which to build a bank of dependable data based on controlled research and application experience.

Titanium's most attractive feature is its very high strength-to-weight ratio. At 57% the weight of steel and with strength properties matching those of heat treated alloy steels, titanium alloys are ideally suited for aerospace, jet engine, and missile construction. Their most serious drawback however stems from a prohibitive cost that discourages the choice of titanium alloy for fastener application other than the few for which there is no practical alternative.

Titanium alloys exhibit superior corrosion resistant behavior in a difficult environment such as that found in a chemical processing plant. Other advantages include reasonably good high and low temperature properties, nonmagnetic behavior, low thermal conductivity, and low coefficients of thermal expansion. On the negative side, fastener fabrication is difficult. Threaded fasteners made of titanium alloy have a tendency to gall and seize during installation and tightening which necessitates careful lubrication. There is also some contradicting evidence that titanium alloys may have a susceptibility to stress corrosion cracking at moderately elevated temperatures. However this problem can be alleviated through special processing.

Unalloyed titanium is non-heat-treatable so that very few fasteners are made of "pure" titanium. There are numerous titanium alloys which are however proprietary developments. Through research a few have surfaced as being the most suitable for threaded fastener manufacture. The workhorse here is the alloy Ti-6A1-4V. Fasteners made from this material have minimum tensile strength of 135 ksi (930 MPa), 125 ksi (860 MPa) yield strength at 0.2% offset, 96 ksi (660 MPa) shear strength, 10% elongation, and a hardness of Rockwell C 30/36. Ti-4A1-4Mn alloy has good creep resistance, stability, and forges well. The relevant strength properties are similar to those of Ti-6A1-4V but with a slightly smaller ductility. Ti-1A1-8V-5Fe alloy delivers a tensile strength of about 200 ksi (1380 MPa). Because of the high strength-to-weight ratio, fasteners of this alloy produce the same holding power as steel fasteners of the same weight and a tensile strength of 350 ksi (2400 MPa). Two other alloys sometimes used in fastener manufacture are Ti-6A1-12Zr and Ti-6A1-6V-2Sn.

30

Super Alloys

The super alloys are a special group of metallic materials which have performance capabilities extending far beyond those of the more commonly used fastener materials. Qualifying alloys are those with exceptionally high strength properties, stability to perform in extremely high temperatures as well as cryogenic applications. Most of these materials are proprietary and trade named. The fasteners made from such alloys are very expensive and should not be specified casually.

In addition to the more routine mechanical properties associated with threaded fasteners, such as yield, tensile strength, hardness, and ductility, the super alloys introduce the fastener engineer into an entirely new world of properties such as creep, stress rupture, stress relaxation, notch sensitivity, high temperature oxidation, thermal expansion, and thermal fatigue. As this book is primarily concerned with the conventional rather than the exotic applications only the briefest explanation of the specific properties and the process of selection are warranted.

Most metallic materials are stable under stress at moderately elevated temperatures. However, as temperature increases the material deforms and begins to creep. *Creep* is the deformation that occurs over a period of time when a constant stress and temperature are applied. Rate of creep then is a

135

function of stress and temperature. An increase in either causes an increase in the rate. *Stress rupture* on the other hand is a measure of the time it takes for a material to rupture when the applied stress and temperature are kept constant. Creep is concerned with the dimensional changes while stress rupture is concerned with fracture. Together they control the ability of a material to survive in a stress and time domain.

In the case of mechanical joints working at elevated temperatures, the stressed length between the bearing surfaces of a bolt head and the nut can increase and gradually lose fastener preload due to creep. This characteristic is known as *stress relaxation* and while it relates to creep the difference is that creep is constant stress-changing strain while stress relaxation is constant strain-chainging stress. Stress relaxation is a serious consideration in the design of joints intended for high temperature environments because loss of a fastener preload introduces a distinct possibility of joint loosening and fastener fatigue.

In subzero atmospheres most materials lose their ductile character and assume a brittle behavior. *Notch sensitivity* defines the degree a material is weakened by the presence of a notch and it aggravates brittle behavior. Fasteners with their changing cross-sections due to the head and thread discontinuities must have notches and the severity of sharpness of these dimensional changes directly affects the fastener's ability to absorb stress punishment. Consequently, for low temperature service, design attention usually given to metallurgical considerations should be extended to fastener dimensions and geometry.

High temperature oxidation is the process of corrosion at elevated temperatures. Depending on the fastener environment the designer can be faced with a real challenge to find the means to combat high temperature and corrosion effects simultaneously.

All metals expand when heated. In any joint structured of different materials and subjected to high temperature service, some provision should be made to accommodate dimensional change due to different coefficients of *thermal expansion*. If the joint components are close fitting and no allowance for size change is provided, thermal stresses can develop and conceivably cause a failure.

Thermal fatigue denotes cyclic thermal stressing and it is related to the differentials in the thermal expansion, magnitude of the temperature change and the frequency of thermal cycles.

Here are a few of the many super alloys suitable for fastener manufacture. H-11, Inconel 718, Vascojet M-A, and MP35N are used for fasteners requiring tensile strengths of 220 ksi (1510 MPa) and much higher. A-286 (a real workhorse), Discaloy, 19-9DL, W-545, Unitemp 212, and Greek Ascoloy are high strength austenitic iron-base alloys. Specifications for the first four are covered in ASTM A437. Rene 41, Inconel X-750, the Hastelloys, Udimet 700, M-252, Inconel 718, Waspaloy, and Astroloy are austenitic nickel-base alloys with excellent high temperature properties. Inconel 718, A-286, and Unitemp 212 exhibit superior performance at extremely low temperatures. Chemical composition, mechanical properties and performance data on most of these and similar super alloys may be found in SAE Report J467.

31

Fasteners of Nonmetallic Materials

Probably less than 1% of all mechanical fasteners are made of a nonmetallic material. Nevertheless, plastics provide certain combinations of performance characteristics which no metallic material can duplicate. Plastics are light weight and corrosion resistant. They also have excellent thermal and electrical insulating properties, and can be easily colored for matching or coding. To their detriment however, plastics have low strength and cannot tolerate even modestly elevated temperatures. Furthermore many of the plastics embrittle rather quickly in moderately low temperatures.

Nylon-6/6 is perhaps the most widely used plastic for fasteners. Its strength properties and the ability to be torqued, are good. This plastic is relatively tough and has resistance to creep and fatigue. Further attributes include excellent elasticity and resiliency to rebound to its original shape following a deformation. The material is reasonably impervious to corrosive attack by most chemical solvents and it is self-extinguishing if set afire. Unfortunately, nylon has a relatively high moisture absorption which affects the dimensional stability. In a particularly hot, dry, and sunny outdoor exposure nylon has a serious tendency to embrittle.

Fasteners of high density *polyethylene* have medium strength properties compared to nylon. However this plastic has outstanding dielectric properties, can be easily molded, and, most importantly, it is inexpensive. *Polystyrene* is also of low cost and has good dielectric properties although it becomes brittle and will crack when impacted or loaded in shear. *Polycarbonate* fasteners have excellent toughness, and can accept impact and shock loading. Their mechanical and dielectric properties are generally attractive. Polycarbonate is dimensionally stable and offers good optical properties, including transparency. Unfortunately, optical properties may deteriorate rather quickly if exposed to direct sunlight.

The most heat resistant plastics are the *fluorocarbons.* For instance fasteners made of tetrafluoroethylene (TFE) can perform in exposures up to about 400°F. TFE fasteners display corrosion resistant properties superior to any other organic material and their ability to survive extended outdoor weathering is excellent. Unfortunately, their strength properties are relatively low while their cost remains extremely high.

Acrylic fasteners have modest use in applications requiring transparency, although their other engineering properties are generally mediocre. The *acetals* are suitable for fasteners requiring extreme dimensional stability, and low moisture absorption. Acetal also has a low coefficient of friction which helps in the situations where easy sliding or contact lubricity is of importance.

Flexible vinyl recognized for good elasticity, is resistant to corrosive attack by most chemicals, and has excellent flexibility or use as a grommet-type underhead sealer. *Rigid vinyl, PVC,* has superior outdoor weathering characteristics and corrosion resistance, particularly in contact with acids.

Relatively few companies in the United States specialize in the manufacture of nonmetallic fasteners. However, those that do are well experienced and have extensive technical literature available for distribution on request. The engineer contemplating the possible use of plastic fasteners is well advised to take advantage of their services.

32

Behavior of Fasteners

It has been stated on many occasions that despite considerable progress in
fabrication techniques, experimentation, and design, our knowledge of
threaded-fastener behavior continues to be rather incomplete. One of the
reasons for this is that in many engineering circles design of bolts is considered to be a mundane topic and there seems to be little incentive for
developing additional information. Our own experience, design handbooks,
and regulatory documents are used to guide us in matters of the final choice
of a particular fastener with little scope left for enhancing our basic knowledge of fastener behavior. One can argue that this approach is practical
and economical in many design situations, and therefore, there is very little
need for getting deeper into theory.

Various chapters of this book have so far been devoted to the extremely
important areas of materials selection, fastener grades, and industrial standards. The remaining portions of this volume relate to mechanical behavior,
design, and working formulas related to fastener selection.* Bearing in mind
the practical approach to fastener engineering, only the simplest concepts
and formulas will be presented.

*For more complete treatment see the book by J. H. Bickford, "An Introduction to the
Design and Behavior of Bolted Joints," Marcel Dekker, Inc., New York and Basel, 1981.

141

The conventional reference to bolt geometry should be extended to all threaded fasteners such as studs and machine screws. The fastener behaves essentially as a clamping device as soon as the nut or head is turned to produce a small amount of stretch. The stretch creates stress and strain in the axial direction while the frictional resistance of the nut and thread surfaces causes a torsional effect resulting in a torsional stress. Furthermore, if the initial alignment of joint interfaces is not exactly parallel a certain amount of fastener bending cannot be avoided. It becomes gradually apparent that our original mundane problem becomes more complex, particularly as the joint surfaces and the gasket materials compress and then relax under the initial preload.

It may be of interest to consider here what happens when we tighten a bolt. In order to be effective as a clamp, the bolt is stretched. The process of stretch generates axial forces which hold the nut in place. When we stop turning the head of the bolt some of the initial stresses generated by stretching and twisting are relaxed. This relaxation can be much greater when relatively soft gaskets are used. The initial clamping force is usually referred to as preload. If we develop too much preload the bolt or the parts held together may be damaged. When the amount of preload is not sufficient, the joint members can slip or separate slightly with respect to each other. The joints which slip can create leaks, vibration, and accelerated conditions of fatigue. Joint separation due to insufficient preload can be particularly difficult to control because you do not have to, physically, separate flange faces to create a leak. Sometimes a mere reduction of interface pressure can accomplish this.

Experience shows that the amount of the initial fastener preload is very important in mechanical joint design. Specifications and standards developed in various countries are essentially aimed at reducing the cost and enhancement of joint reliability. These two characteristics are not necessarily compatible because of the many variables involved, and the empirical nature of design uncertainties. The statistics of fastener behavior is in itself a complex problem in mathematics and provides little help in comprehending the acceptable confidence limits for a particular joint design. It can be stated that the amount of preload depends on various geometrical and physical features, the characterization of which require a significant number of experiments. Unfortunately, in the real world we can only afford to test a few samples. It should also be recalled that the manner in which the results from individual tests are grouped together follows the so-called normal distribution defined by a bell-shaped curve. However, not all distributions will be found to be normal, making a study of a particular effect more difficult. The approach

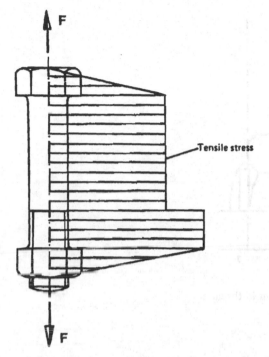

Figure 8 Simple stress pattern for a fastener in tension.

based on the analysis of test results and the calculation of confidence limits falls in the area of a probabilistic design. The material which follows, however, only outlines the deterministic approach to a design solution using the conventional formulas and criteria. In other words, the designer can calculate, say, a specific torque, preload, or a stress level in a particular fastener on the assumption of certain dimensions and mechanical properties available.

The stress distribution in a fastener is the result of complex structural behavior. This complexity alone has been sufficient to intrigue a great many theoreticians and experienced design engineers in all walks of industry and educational institutions. Even though the bolt has been subjected to a pure tension, there are compressive effects caused by Poisson's ratio and the compressive forces in the nut and bolt-head caused by the tightening. In addition, there are localized discontinuities affecting the stress distribution. The

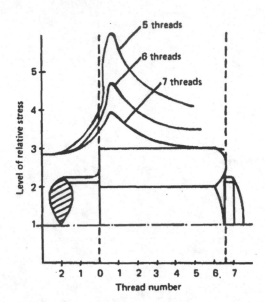

Figure 9   Example of stress variation in threads.

more obvious areas of significant gradients, of course, are the fillets and
thread profiles difficult to treat analytically and experimentally. A simplis-
tic idea of the stress pattern in a threaded fastener plotted along the central
axis is shown in Fig. 8.

   Essentially, the stress distribution given in Fig. 8 is based on the nominal
stress value F/A where F is the axial load and A is the cross-sectional area of
the fastener at the particular point under consideration. This simple model of
stress behavior is reasonably justified for relatively long bolt shanks where the
bolt length to diameter ratios are greater than about 4. In practice, the majority of fasteners tend to be rather short and stubby. Under such conditions,
even the F/A stress cannot be very uniform across the shank diameter. As
to the actual picture of stress positioned away from the central axis, even
sophisticated finite element calculations fall short of predicting the correct
stress gradients. This is particularly true at the surface of the fastener.

   Several aspects of stress behavior were already discussed in Chapter 7. It
can be stated here, in general, that the complexity of the problem increases

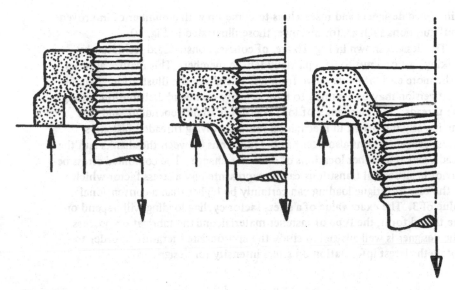

Figure 10 Improved nut configurations for load transfer.

as one considers the variation of load transfer across the thread array. It is important to learn that peak stresses in nut or fastener threads fall-off in a nonlinear fashion with the number of threads as illustrated in Fig. 9.

The stress behavior given in Fig. 9 is brought about by the two main factors. One is due to the nonlinear response of the consecutive threads while the other is concerned with the stress intensity at the bottom of each thread where the actual radius of curvature is extremely small. It appears that the first two or three threads carry the lion's share of the bolt load so that adding more threads and making longer nuts is not going to ease the problem of peak stresses significantly. It should also be recalled that the axial load F (Fig. 8) tends to increase the thread pitch for the fastener with the corresponding decrease of the pitch of the nut. This situation complicates the mechanism of load transfer further, consistent with the theory that the thread closest to the bearing surface of the nut must be overloaded. Experience shows that in the conventional design of a nut having six or more threads, the thread next to the bearing nut surface can carry as much as 35% of the total load. This, of course, leads to the various modes of thread stripping as discussed in Chapter 7, and the problem of uneven load transfer has

stimulated designers and researchers to come up with a number of improved
configurations such as, for instance, those illustrated in Fig. 10.

The designs shown in Fig. 10 are, of course, nonstandard, achieved by
making part of a nut to be a partially tensile member. This feature appears
to be more evident as we move from left to right in the illustration. The far
right version may be assumed to have a nearly even load distribution among
the threads. The mechanics of load transfer is strain dependent, and it is
particularly important in mechanical joints involving threaded fasteners.
Stress concentrations also occur at the junctions between the shanks and the
heads, and at all other locations of diameter change. The combined stress behavior in structural transitions can be represented by a stress factor which
in the case of fatigue loading can certainly be higher than a conventional
value of 3. The exact value of a stress factor cycling loading will depend on
the thread form, the type of fastener material, and the fabrication process.
The designer is well advised to study the appropriate literature in order to
obtain the latest information on stress intensity for design.

33

Stresses in Fasteners

It may be instructive here to ponder those elementary cases where individual fasteners behave as conventional structural beams in tension, bending, or shear. As stated in Chapter 1, the elementary case of a fastener shown in Fig. 2 may have a variable cross-section. The total change in length under the axial load F follows then from the application of the well-known Hooke's law, which, in the analysis of the fastener given in Fig. 2, yields:

$$\Delta L = \frac{F}{E}\left(\frac{L_1}{A_1} + \frac{L_2}{A_2} + \frac{L_3}{A_3} + \frac{L_4}{A_4} + \frac{L_5}{A_5}\right) \tag{1}$$

In the formula given by Eq. (1) E denotes the conventional modulus of elasticity expressed in psi (using English units) while L and A (values as indicated in Fig. 2) are the length and cross-sectional dimensions in linear and square inches, respectively. For a constant cross-section A and the total length L equal to $L_1 + L_2 + L_3 + L_4 + L_5$, Eq. (1) reduces to the familiar elementary expression:

$$\Delta L = \frac{FL}{AE} \tag{2}$$

This formula leads to the important definition of a spring constant, K_f, often used in the analysis of mechanical joints held by fasteners:

$$K_f = \frac{F}{\Delta L} = \frac{AE}{L} \tag{3}$$

Although the calculations of a spring constant using Eqs. (1-3) appear to be simple, the exact dimensions, mechanical properties, and the fabrication techniques may be of some importance in the measurement of bolt stretch (ΔL) or a spring constant as an indication of preload. The nominal axial stress in tension is, of couse;

$$S_t = \frac{4F}{\pi D^2} \tag{4}$$

Here D denotes the nominal fastener diameter, so that the corresponding cross-sectional area becomes:

$$A = \frac{\pi D^2}{4} \tag{5}$$

The general symbol A refers here only to the nominal cross-section. The actual areas of interest in stress calculations can be estimated on the basis of nominal, major diameter of thread, or root diameter. These are usually called tensile stress and thread root areas, respectively. In inch series these parameters are expressed in terms of the nominal diameter D and the number of threads per inch is denoted by n, as shown in Chapter 7. Similarly, these areas are given as a function of D and p for the metric series, where thread pitch is p.

For a double shear joint such as that illustrated previously in Fig. 3, there is one shank and one threaded cross-section. For the conventional fastener given in Fig. 3, the formula for the maximum shear load can be stated as follows:

$$Q_{max} = \left(\frac{\pi D^2}{2} - \frac{1.5D}{n} + \frac{0.75}{n^2} \right) \tau_{max} \tag{6}$$

There may be a number of different configurations where either one or several shear planes through the fastener can be involved in a clamped condition. Because of the corrections of stress areas suggested in Chapter 7, Eq.

(6) appears to be having a dimensional problem in some of the terms. In other words, the shear load can only be obtained by multiplying the shear strength τ_{max} by a term containing D^2. Incidentally the first term in Eq. (6) is by far the largest and the apparent problem with the dimensions exists only in the inch series of fasteners. In practice all the formulas for stress and root areas yield acceptable results. The actual factor of safety in shear for the configuration given in Fig. 3 can be obtained by dividing Q_{max} by the applied load F. In this discussion the external load is defined by F for the transverse as well as axial load applications.

There are usually two types of shear that a fastener can carry. The first one is of the direct, transverse type described by Eq. (6). The second type is due to the external torsional moment M_{to} applied to the fastener for the purpose of overcoming frictional, geometrical and tensile constraints of the joint. The tensile constraint may be considered to be caused by stretching the shank. The geometrical constraint relates to the form of the thread while the frictional resistance comes from the interaction of thread surfaces under clamped conditions. Assuming that the effective radius of contact between the threads is not substantially different from D/2, the external torsional moment can be defined as:

$$M_{to} = \frac{F'}{2} \frac{p}{\pi} + \frac{\mu D}{\cos \beta} \tag{7}$$

Here F' stands for the axial force under torqued conditions which may or may not be equal to the applied external load. The coefficient of friction is denoted by μ, while β is the half angle of the thread profile, which for international thread design, is 30°. If the shank now behaves as a cylinder in torsion, then the maximum torsional stress at the outermost fiber of the shank may be on the order of

$$S_{to} = \frac{0.81F'(p \cos \beta + \pi\mu D)}{D^3 \cos \beta} \tag{8}$$

If the misalignment of the flange surfaces held in place by the fasteners prevents a purely axial response in these fasteners, then it is necessary to postulate the existence of an external bending moment M_b caused by the flange misalignment. Assuming next that each fastener in the joint is bent to a radius of curvature R, then the external bending moment corresponding to this curvature is

$$M_b = \frac{\pi E D^4}{64R} \tag{9}$$

The resultant bending stress becomes

$$S_b = \frac{ED}{2R} \tag{10}$$

It may be of interest to note here that, essentially, Eq. (10) represents the well-known Hooke's law if we rewrite the above expression as

$$\frac{D}{2R} = \frac{\Delta L}{L} = \frac{S_b}{E} \tag{11}$$

It is quite probable that the radius of curvature R due to misalignment can be very large compared to the fastener diameter D. Hence the term D/2R must be quite small and probably on the order of a typical $\Delta L/L$ ratio featured in Eqs. (2) and (3). Dimensionally D/2R represents strain as implied by Eqs. (2), (3), (10), and (11).

So far we have attempted to treat a given fastener as an individual structural beam subjected to various loading conditions such as tension, shear, torsion, or bending. It soon becomes obvious, however, that the combinations of loading should lead to progressively more complex cases of fastener behavior.

Experience shows that in the majority of real applications fasteners are subjected to a combined stress sytem involving tension, shear, torsion, and bending. Unfortunately, the right choice of a load combination, geometry, and physical properties is practically never available for a reliable approach to the optimum design. While our primary interest may well be to determine the tensile rather than shear strength of a given fastener, it should be expected that the tensile strength is reduced when torsion or shear loads are present. This condition develops as soon as the torque is applied. There is an inevitable interaction between the axial and torsional loading in overcoming the frictional resistance of a fastener.

It may well be of interest at this stage of our review to summarize a few basic principles. A typical threaded fastener is designed to be tightened by twisting the nut with respect to the bolt through the application of a specified torque. During this process the nut turns and the bolt (fastener) stretches to create the required preload. Our control of this operation can be achieved by either the amount of the applied torque, turn of the nut, stretch of the shank, or some combination of these three basic actions. In general, the control of preload by means of torque and/or turn techniques is likely to be the easiest and least costly. The only problem is that the relationship between torque or turn and the required preload is not always easy to predict.

34

Formulas for Torque and Preload

The design theory of fasteners has, over the years, been influenced by experimental work aimed at defining yield, proof, and failure criteria for a single fastener. The main task was to estimate the tensile strength of the threaded portion of the shank, which was difficult to do with the elementary equations such as, for instance, those discussed briefly in Chapter 33. Over the years a large number of fasteners made from well-characterized materials were tested and manufacturers were required to repeat many of these tests in order to build-up technical confidence in the product. While a great many test results have also been published, our need for constant vigilance of quality and improved calculational procedures has not gone away. One important practical result, however, has remained essentially unchanged, indicating that the yield load on the fastener in pounds divided by the stress area results (essentially) in a theoretical stress at yield. At the same time many actual applications show that the fastener seldom sees the yield stress. The basic requirement in fastener design is not necessarily to predict the exact yield stress but to select the appropriate torque M_t and the required preload F_i with a rational margin of design safety. The comparison of F_i values with published strength data for a particular fastener size offers, at this time, the most reliable design alternative.

Preliminary formulas, such as those given by Eqs. (7) and (8), are based on the work of Bickford and represent the torsional response of the fastener. The total torque required to overcome geometrical and frictional restraints of the joint can be calculated from the following equation:

$$M_t = F_i^r \left(\frac{\cos \beta \tan \alpha + \mu}{\cos \beta - \mu \tan \alpha} + \frac{\mu R_0}{r} \right) \tag{12}$$

In the above formula α defines the helix angle of the thread, r is the mean radius of the thread while R_0 stands for the effective radius of the frictional area under the nut. The remaining symbols are the same as those employed in Chapter 33.

Experiments as well as theoretical analysis suggest that there is a linear relationship between the applied torque and the developed preload for a given fastener geometry. That is,

Torque = Preload X Constant

In the case of Eq. (12) the constant depends on r, R_0, α, β, and μ. Various formulas have been, in the past, derived in terms of these and similar parameters in order to have a method for estimating the value of the constant. One can recognize three separate terms inside the brackets of Eq. (12) which may be characterized as three reaction torques caused by the inclined planes of threads, friction between nut and bolt threads, and the frictional resistance between the face of the nut and the joint surface.

The well-known practical formula for approximating the amount of tightening torque is:

$$M_t = \frac{DF_i}{5} \tag{13}$$

It should be noted that this expression is extremely simple indicating that the parameters α and β can be eliminated by adopting some typical numerical values. For instance, when α is approaching zero, substituting $\mu = 0.15$, making β equal to $30°$, and $R_0/r = 1.4$, Eq. (12) transforms into Eq. (13). This approximation seems reasonable when we realize that the helix angle α is indeed very small in practice, and that $\beta = 30°$ corresponds to the half-angle of a thread profile accepted by international standards. Also the assumption of $\mu = 0.15$ is compatible, for instance, with the frictional factor for a typical dry surface of a steel fastener. It may further be observed that in a more general sense the design equation defining the torque can be stated as follows:

$$M_t = KDF_i \qquad\qquad (14)$$

The factor K in Eq. (14) combines the effect of the frictional surfaces, geometry of thread profile, and the characteristics of industrial lubricants. The average K values for steel fasteners are given in Table 18.

In order to estimate the torque value in inch-pounds, the parameters D and F_i in Eq. (14) should be given in inches and pounds, respectively. For instance, calculate the torque required to develop a preload of 5,000 pounds if the fastener diameter is 0.75 inches and the lubricant is moly grease. Since Table 18 gives K = 0.14, utilizing Eq. (14), yields:

$$M_t = 0.14 \times 0.75 \times 5,000 = 525 \text{ lb-in.}$$

The elementary and convenient form of Eq. (14) shows that the preload F_i is quite sensitive to friction which constitutes the main portion of factor K. For a given torque and fastener diameter the preload is inversely proportional to K. Theoretically, when K is rather small the amount of preload can be very high. In practice factor K is not expected to approach zero but it is well to keep in mind the basic relation which follows directly from Eq. (14),

$$F_i = \frac{M_t}{KD}$$

Table 18 Average K Factors for Steel Fasteners

Industrial lubricant	K
As received, steel	0.20
As received, cad plate	0.19
Phosphate and oil	0.19
Parkerized and oiled	0.18
Moly grease	0.14
Fel-Pro 65A	0.13
Petroleum, light oils	0.12

The values of K may be different for different fastener materials and the designer is advised to consult the manufacturer about the recommended lubricants and preload factors similar to those given in Table 18. Where the torque and preload are not critical in a particular design the approximate formula given by Eq. (13) is quite satisfactory.

The calculation of stresses in a threaded portion of a fastener is not a simple matter and it is fortunate that such a detailed stress analysis is seldom required. It can also be stated that the combined stress S, induced by tension and torsion, depends primarily on the fastener diameter and frictional effects on contact surfaces. The plot in Fig. 11 represents the dimensionless parameters SD^2/F_i expressed as a function of the coefficient of friction μ, which is not quite the same as the factor K given in Table 18 although the relevant numerical values are at times reasonably close.

Taking, for instance, $\mu = 0.15$, the dimensionless parameter SD^2/F from Fig. 11 is equal to about 1.53. Hence using previous numerical values, one obtains:

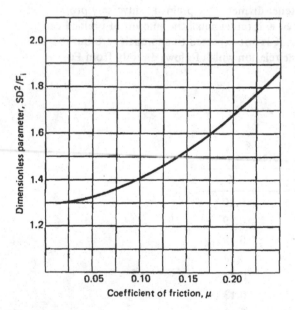

Figure 11   Effect of friction on fastener stress.

$$S = \frac{1.53 \times 5,000}{0.75^2} = 13,600 \text{ psi}$$

Although there is definitely a close relationship between the K and μ values, both parameters can be discussed separately. The average K values given in Table 18 can be extended to 0.08 and 0.27 as reported in industry. The value of 0.20 reflected in Eq. (13) is, of course, an approximation applicable to fasteners and nuts having a rather plain and uplated finish without lubrication. It is also generally believed that about 50% of the wrenching torque may be used-up to overcome the friction between the nut and the bearing surface with some additional 40% allocated to overcome the sliding resistance between the threads. As indicated previously, only 10% of the torque is utilized in developing the required preload. In a more detailed analysis the torque and preload factors K can be expressed as follows:

$$K = K_1 + K_2 + K_3 \tag{16}$$

In this formula, K_1 relates to the bearing friction under the nut, K_2 defines the frictional effect on the thread surfaces, and K_3 is a geometrical parameter depending on the helix angle of the thread and other fastener proportions. It appears that K_3 represents roughly the useful portion of the torque.

As stated elsewhere in this chapter the coefficient of friction appearing in Eqs. (7), (8), and (12), as well as Fig. 11. is generally lower than the corresponding value of the torque and preload factor K. Some of the representative values of μ are given in Table 19.

It should be recalled again that for a given factor K and fastener diameter D there is a linear relationship between the applied torque and the preload as shown by Eq. (14). Experience indicates in general that such a linear relationship can indeed be confirmed experimentally. The departure from the straight line becomes evident only when either the fastener or the joint material begin to yield. While a remarkable correlation exists between the simple torque formula and the experiment, our understanding of the frictional effects continues to be rather diappointing. It does not take very long to list a dozen variables affecting the manner according to which friction in a fastener assembly works.

The parameter K, referred to in this chapter as the torque and preload factor, is often called the "torque coefficient" or "nut factor." The values of K can only be determined experimentally for each new joint design. Thread lubrication can help to improve repeatability of test results for this

Table 19 Friction Factors of μ

Lubricant	Range of μ
Red lead, graphite mineral oil	0.078–0.110
Red lead, graphite, machine oil	0.055–0.065
Graphite, mineral oil	0.035–0.060
Graphite, machine oil	0.035–0.055
Molokote G	0.030–0.075
Fel-Pro C5A	0.045–0.080
Crane compound 425A	0.070–0.090
Grafo 360	0.100–0.115
Neolube	0.030–0.090

purpose. However, other influences such as geometry, tool accuracy, relaxation, design quality, operator skills, and human error introduce significant variations and overshadow our ability to correctly determine the value of K or μ.

35

Working Load Criteria

When only a rough estimate of a fastener preload F_i is required the calculation can be made using the following approximation:

$$F_i = 5 \frac{M_t}{D} \tag{17}$$

Assuming that the torque coefficient K is known with somewhat better accuracy, Eq. (15) should be used in the design. However, it is well to keep in mind that the fastener working load is not the same as the preload and for many good reasons.

Experience shows that one of the primary considerations in a mechanical joint design is the selection of the ratio of the working load to the preload. This chapter attempts to illustrate some of the basic rules and design philosophy related to the working load criteria.

In a practical design case we should start with the required working load, fastener material, and the estimated load carrying capacity of the joint in order to find a factor of safety for our design. This is not very easy, however, because the strength of the assembled joint is affected by the many factors and interactions which, given the state of the art, are still extremely difficult to assess. Such important variables as the torque coefficient, spring constant of the joint, potential leak rate of the gasket or subtle effects of load eccentricity of the fastener tension contribute to the complexity of design. Many data points have to be established experimentally before our engineering judgment can be used in the final analysis.

If for instance a fastener is selected in such a way that its yield strength is four times the required working load, it does not automatically mean that the design will have a safety factor of 4. This can only be true if the relevant clamping load is four times the working load and our joint has to be perfectly rigid. It can also be shown that the so-called flexible joint can only be tightened to the level of the working load.

In the majority of practical design cases the ratios of the working loads to the preloads range between 0.3 and 0.8. It appears that as there ratios increase the margin between the working load and the preload decreases. It also follows that the use of a more efficient design requires more reliable fastener information, which is not easy to get. From a safety point of view a wide margin between the external load and the preload is desirable which, however, implies a relatively conservative design and a high cost.

One of the cardinal design rules states that in a reliable joint the working load should not be permitted to exceed the fastener preload. This practice is especially important in the case of fatigue applications. Then, as long as our fastener does not experience any appreciable stress variation, there can be no failure in fatigue regardless of the number of load cycles. Again, this rule loses its meaning when the joint displays considerable compressibility so that the presence of even moderate stress may be high enough to cause fatigue failure regardless of the initial preload.

As indicated previously the highest stress is found at the first thread inside the nut, and this is the most critical location of fastener failure due to fatigue. Similar critical points can be found at the junction of the head and the shank or at the thread runout. It appears in general that the best fatigue resistance is obtained when the rigid members of the joint are held together by elastic fasteners. High preload also signifies a good deal of assurance that the joint can withstand the intended higher static and cyclic loads which would not be able to pull the joint apart.

While a reasonably high preload is beneficial to joint design, it is well to keep in mind the possibility that a limited amount of plastic elongation of the fastener can occur. Such an elongation of course may result in some loosening of the joint. Hence, whenever practical, the peak fastener stresses should

be kept below the elastic limit. Such stresses can often be predicted with the
help of a simple expression (SD^2/F) and the design curve shown in Fig. 11,
by making $F = F_i$. Conversely, the allowable preload can be estimated when
$S = S_y$, where S_y denotes the yield strength of the fastener material. For instance, taking $\mu = 0.2$, $D = 0.75$ in., and $S_y = 130,000$ psi from Table 3, we
obtain:

$$\frac{S_y D^2}{F_i} = 1.68$$

from which

$$F_i = \frac{130,000 \times 0.75^2}{1.68} = 43,527 \text{ lbs}$$

Unfortunately, joint loosening may sometimes be unavoidable despite
our efforts to calculate and specify the limiting preload. It can however be
minimized by tightening larger fasteners to somewhat higher values of the
initial tension. The practice of the appropriate balance between the high preload and the minimum joint loosening is, however, a matter of experience and
delicate compromise.

The determination of the working loads depends on the choice of the factors of safety, yield strength of the fastener material, and the actual parameters of joint stiffness. If the assumption can be made that the joint components do not separate as a result of the application of the external working
load, then the decrease in the deformation of the connected members must
be equal to the increase in the deformation of the fastener. This basic relation permits setting up of the design formulas for the calculation of the
working load in terms of the relevant spring constants and the fastener preload.

It is clear from the preliminary considerations of torque and preload criteria that the coefficient of friction, as usual, is the main culprit. In taking a
closer look at this problem we will indeed be amazed to discover that about
30 to 40 variables may be involved in the process of predicting the effects of
friction in a threaded fastener. We will not attempt to name them all but
certainly such variables as hardness, surface finish, types of lubricants, con-
tamination of surfaces, temperature, hole clearance, fit between threads, sur-

face pressures, and other effects will be on the list. Many of these variables cannot be reliably controlled without appreciable cost and effort. Others are simply beyond our capability to understand and predict.

In concentrating on friction as our major villain we must not underestimate the effect of geometry. There are unexpected variations in the pitch of the threads, half-angle of thread, of perpendicularity of the faces of the nut in relation to the thread axis. These variations can be further accentuated by local plastic deformations developed in the areas of severe stress concentration. In addition the true relationship between the torque, say, applied to the bolt and the resultant perload is further complicated by the energy losses due to bolt twist, bending of shank, or nut deformation. One can imagine an extreme case where the threads gall and seize so that the entire torque is lost in thread friction. Under such conditions, of course, the energy converted to preload approaches zero. Fortunately, in the real world, even a small portion of energy can be used up to develop some preload since the condition of infinite friction is not possible.

36

Spring Constants

It is important to realize that our knowledge of fastener behavior would be totally unrealistic and incomplete without the proper appreciation of joint behavior and those variables which affect our choice of fastener design and load limitations. The mechanical connections can only be characterized with the aid of the stiffness properties, which in the majority of cases, are referred to in design literature as spring constants. The material in this chapter includes the basic concepts and definitions of spring constants; they are denoted as

K_f Spring constant of fastener (bolt)
K_c Spring constant of the joint
K_g Spring constant of gasket
K_j Spring constant of connected plates or flanges

Containment of various liquids or gases presents a very serious problem both from an economic and safety standpoint. The requirement of joint tightness is difficult because of the extra measure of accuracy and control effort placed on joint and fastener design. However, the most perplexing uncertainty comes from gasket behavior. Despite the progress made in many branches of industry our knowledge of gasket behavior is still archaic and therefore, nobody, as yet, has produced a truly leak-free joint. We simply have to live with the problem and design gasketed joints without complete theoretical answers.

The control of leaks requires sufficient contact pressure, often called gasket-stress, which should be maintained even after the gasket and bolt material relax. The contact pressure is, of course, developed as the bolts are tightened. It is important to note that nonuniform tightening can lead to uneven distortion and opening of leak paths. Rough or damaged flange surfaces can further increase the leak rates.

The gasket can be regarded as a spring in series with the bolt, nut, washer, and flange springs.

It has been generally observed that the spring constant K_g is usually smaller than the spring constants K_f or K_j. However K_g is nonlinear and depends on the force-deflection behavior of the gasket. Industrial data usually shows the gasket characteristics as plots of compressive stress versus deflection. The problem is that it is difficult to represent the behavior of a gasket by a single spring constant or a simple design formula. Gasket material is not totally elastic since it exhibits hysteresis and permanent set, and the coefficient of thermal expansion of the gasket material is usually quite different from that of the fastener or the flange.

The two important forces acting in a mechanical joint held together by fasteners are F_i and F_e, denoting the initial preload in tension and the external load respectively. If we assume that the joint has not separated under the applied load F_e, the decrease in the deformation of the connected parts must be equal to the increase of the fastener extension. For a metal-to-metal, rigid contact in a typical joint, there must be certain simultaneous axial tensions and frictional effects such as those shown in Fig. 12.

The actual load carried by the fastener in a mechanical joint can be expressed in terms of the applied external load and the initial preload as follows:

$$F = F_i + CF_e \tag{18}$$

The formula given by Eq. (18) is essential to understanding the behavior of fasteners in a mechanical joint where C stands for the overall stiffness coefficient, or a spring constant of the assembly. In this notation the overall spring constant can be stated in the followng manner:

$$C = \frac{K_f}{K_f + K_c} \tag{19}$$

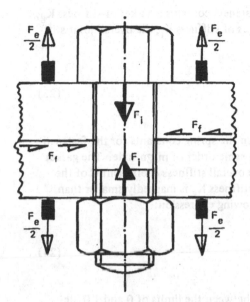

Figure 12 Equilibrium of forces in a mechanical joint.

Quick reference to Fig. 12 shows that the parameter C does not include any algebraic terms related to frictional forces, F_f, acting in the transverse direction. The important design criteria can be defined with the aid of the relation of the parameter C to the fastener spring constant K_f. When a relatively soft fastener holds rigid components of a mechanical joint together ($K_c \gg K_f$), the following condition will exist:

$$F = F_i \tag{20}$$

For the case when all the joint components and fasteners have essentially equal spring constants ($K_c = K_f$), we have:

$$F = F_i + 0.5F_e \tag{21}$$

Finally, when a very rigid fastener holds a relatively soft joint assembly ($K_c \ll K_f$), the corresponding criterion applies:

$$F = F_i + F_e \tag{22}$$

If a mechanical joint held by the fasteners contains a gasket of stiffness K_g, placed between the two plates or flanges of stiffness K_j, the overall joint stiffness becomes:

$$C = \frac{K_f(K_g + K_j)}{K_fK_j + K_fK_g + K_gK_j} \tag{23}$$

Equation (23) is recommended when the spring constants for the fastened plate and the gasket are roughly of the same order of magnitude. The gasket spring constant influences strongly the overall stiffness coefficient C of the assembly. However, when the gasket stiffness K_g is markedly smaller than K_j, Eq. (23) can be reduced to the following expression:

$$C = \frac{K_f}{K_f + K_g} \tag{24}$$

In this case the value of C can vary between the limits of 0 and 1.0, depending on the actual ratio K_g/K_f. As a general guide the selection of the stiffness parameter C in terms of the fastener and gasket spring constant can be simplified as shown in Table 20. The concepts of stiffness and spring constants are used here interchangeably.

It is recalled here that according to Eq. (23) the stiffness and the spring constant have the following dimensions:

$$\text{in.}^2 \times \frac{\text{lb}}{\text{in.}^2} \times \frac{1}{\text{in.}} = \text{lb/in.}$$

Prior to the application of the external load, the fastener is strained in proportion to the preload F_i. Hence any additional load on a gasketed joint is expected to add to the preload as shown by Eqs. (21) and (22). When F_e is large and when the force on the gasket approaches zero, the joint separates and the fastener feels the entire load F_e.

The foregoing assessment of the joint stiffness is still difficult to make because we struggle with the gasket's nonlinearity. There is virtually no published information about the gasket spring constant and joint relaxation caused by gasket creep. Industry and the engineering profession have at-

Table 20 Guide to Joint Stiffness

Type of joint	$\dfrac{K_g}{K_f + K_g}$	$\dfrac{K_f}{K_f + K_g}$
Soft gasket joint held by studs	0.00	1.00
Soft gasket joint with through fasteners	0.25	0.75
Asbestos gasket joint	0.40	0.60
Soft-copper gasket with long through fasteners	0.50	0.50
Hard-copper gasket with long through fasteners	0.75	0.25
Metal-to-metal rigid joint joint with long through fasteners	1.00	0.00

tempted to bypass this problem by developing two experimental gasket factors called y and m. The y factor relates to the initial gasket stress necessary to seat the gasket in a pressurized system and to mitigate leaks. Since the gasket pressure (or stress) must be larger than the contained pressure in the joint, the ratio of these two pressures is of importance to design. This ratio in gasketed joint design is known as the m factor. Some typical gasket factor values for y and m are given in Table 21.

Design formulas are now available for the calculation of the number and size of bolts for a given preload, using factors m and y. The designer is ad-

Table 21 ASME Gasket Factors

Type of gasket	Gasket factor, m	Miscellaneous seating stress, y (psi)
1/8-in. asbestos with suitable binder	2.0	1,600
Spiral-wound metal (asbestos filled)	2.5–3.0	10,000
Solid and flat gasket (soft aluminum)	3.25	8,800
Elastomer with cottom fabric insert	1.25	400

Source: Adapted from the ASME Boiler and Pressure Vessel Code, Section VIII, Division I, Appendix II, Table UA-49.1.

vised to consult the ASME code for the recommended design procedures. But even the ASME gasket factors should be approached with due caution. Some evidence persists that there can be significant differences between the published factors and experimental results.

37

Prying and Joint Separation

Conventional design of a mechanical joint assumes tht the fasteners and the connected parts follow a linear response. The reason behind the use of elastic equations is their simplicity and frequent lack of suitable design models in more complex situations. In reality the behavior of a working fastener is non-linear. The joint rigidity and fastener tension are a function of the applied loads, geometry, and elastic properties. The corresponding theoretical solutions are difficult to accomplish and the designer has to turn to experiments.

In elementary static analysis the external load is assumed to be in line with the bolt or fastener axis. However, when the external load is applied off to one side of this axis the resultant effect can be described as the prying action responsible for an increase in the fastener load. The basic principle of this action is demonstrated in Fig. 13.

Here F is the external load, Q defines the approximate reaction of the level arm and F_p is the resultant prying force. Taking moments about the edge at Q shows that the additional force on the fastener is proportional to the lever ratio $(a_o + b_o)/a_o$, on the assumption that all the joint members are relatively rigid. Unfortunately this rigid body approach is oversimplified because we have ignored stiffness effects. In reality the problem is highly involved and no general solution is available. As the first approximation

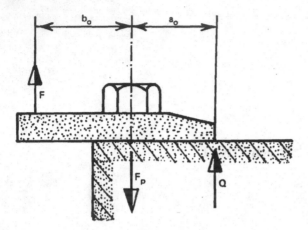

Figure 13 Prying forces equilibrium.

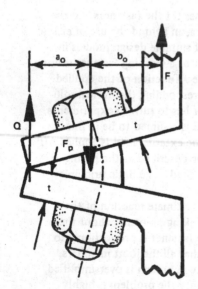

Figure 14 Mechanical joint under prying action.

however, the basic equation of prying action can be postulated with the aid of Fig. 14,

$$F_p = F + Q \tag{25}$$

To date experiments indicate that in the typical tension-type bolted joints the ratio Q/F seldom exceeds 25%. Hence for a moderate value of the external loading the fastener should not see the full impact of the prying action. The exact portion of the external load, however, felt by the fastener will depend on the ratio K_f/K_j until the joint separation occurs. For a rather small ratio of course Eq. (25) can reduce to a very elementary form:

$$F_p = F \tag{26}$$

The parameters of K_f and K_j were defined in Chapter 36.

It also follows that a relatively stiff flange can minimize and even eliminate the effect of prying action entirely. In general the prying forces can be reduced by using the following techniques:

Increasing flange thickness, t
Decreasing dimension, b_o
Increasing dimension, a_o

Addition of the outboard fasteners does not appear to help significantly because the first row of fasteners, nearer to the line of action of F carries the bulk of the load.

While there are a number of mitigating circumstances, the prying action is seldom eliminated entirely. The German practice, for instance, assumes that all the connections should be considered to be eccentrically loaded. In the United States this problem is also recognized and useful design formulas have been developed at times to prevent a joint separation at the point of application of the external load F, as shown in Fig. 15.

An example of a grossly simplified design formula can be given by the following expression.

$$F_p = \frac{F(j - i)}{0.34j + i} \tag{27}$$

The minimum preload F_p does not account for any complex interactions and it is assumed to be sufficiently high to prevent joint separation under the

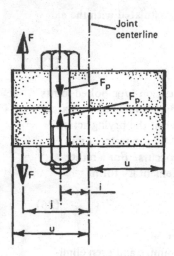

Figure 15   Joint equilibrium for minimum preload to prevent separation.

point of application of the external load F shown in Fig. 15. This is no doubt
the worst location for the onset of joint separation. If we can now replace
the key dimensional quantities i and j by the ratio $\gamma = i/j$, then the design
formula for the critical preload F_p becomes even simpler:

$$\frac{F_p}{F} = \frac{1-\gamma}{0.34+\gamma} \qquad\qquad (28)$$

Here Eqs. (27) and (28) indicate that as the distance from the joint center-
line to the line of the fastener action increases the minimum preload for mitigating joint separation decreases. Hence the mechanism of prying action and
joint separation is interrelated and can be analyzed in terms of similar dimensional parameters as well as the conditions of static equilibrium. The quantities i and j are of course particularly important for this analysis. It should be
added here that simplified formulas given by Eqs. (27) and (28) have been
based on the premise tha the dimension u indicated in Fig. 15 is approximately equal to j. The position of the point of application of the external
load F is often difficult to establish since this load can be created by the various distributions of contact forces resulting from weight, pressure, inertia, or
other effects.

38

Design Calculations

This chapter is intended as a brief illustration of elementary design calculations aimed at the proper selection of fasteners and understanding of their mechanical integrity under various service loads. Adequate strength of individual fasteners, preload criteria, and working external loads are of special importance in assuring satisfactory performance of a mechanical joint based on the size and strength properties of the fasteners selected.

Bearing in mind practical requirements of a design or fabrication office in an industrial environment the material given in this chapter consists of a series of design examples worked out in full detail with the results presented in the English, and where appropriate in SI equivalents. The dual system of units is intended for metric users. U.S. industries, concerned with the mechanical hardware, structural applications, and general machine design are still used to thinking in terms of English units although considerable progress has been made to develop and reconcile the international fastener standards as emphasized in the various chapters of this book dealing with thread designs, fastener proportions, strength grades, and material properties.

The majority of current design calculations are based on the linear formulations and static strength properties although it is well known that a fastener-nut-washer assembly in a mechanical joint involves nonlinearities,

173

plastic effects, and frictional effects which complicate the analysis of fastener behavior. Unfortunately, it will be many years before handbook formulas are replaced by proven, sophisticated computer programs and similar calculational tools. Engineering societies and standards committees recognize this fact and encourage the development of modern design procedures. In the meantime, however, the joint designer is faced with the technical questions related to fastener behavior and materials response which demand his immediate attention. It is hoped that the elementary calculations that follow will, in some small measure, assist the designer in his task. This chapter is in no way a substitute for books devoted entirely to fastener design.

The examples selected utilize the formulas given in Chapters 7 and 32 through 37, as well as the appropriate tables included in this volume.

Design Problem 1
Assuming maximum material condition for the unified screw threads calculate the maximum minor diameter of internal thread in terms of the nominal diameter D and the number n of threads per inch.

Solution: Utilizing the dimensional illustration of Fig. 4, and the material from Chapter 2, we get

$$H = 0.86603p$$

$$p = \frac{1}{n}$$

$$K_{nmax} = D - 2 \times 0.625H$$

and substituting yields:

$$K_{nmax} = D - 2 \times 0.625 \times 0.86603p$$

or

$$K_{nmax} = D - \frac{1.0825}{n}$$

Design Problem 2
Develop a formula for the minimum pitch diameter of external thread E_{smin} utilizing the general proportions of unified screw threads at maximum material condition.

Solution: According to Fig. 4, the distance between the nominal radius and the pitch line for the UN thread is 0.375 H. Hence using the relations from Design Problem 1, we get:

$$E_{smin} = D - 0.75 \, H$$

or

$$E_{smin} = D - \frac{0.6495}{n}$$

Design Problem 3

Calculate the thread stripping areas of external thread AS_s for the inch series fastener having nominal diameter $D = 0.75$ in. and the number of 10 threads per inch corresponding to unified coarse thread (UNC). Assume the length of engaged threaded portion L_e to be equal to the nominal diameter.

Solution: Since threaded pitch is equal to $1/n = 0.1$, K_{nmax} and E_{smin} can be computed from the formulas given in Design Problems 1 and 2.

$$K_{nmax} = 0.75 - 0.10825 = 0.6418 \text{ in.}$$

and

$$E_{smin} = 0.75 - 0.06495 = 0.6851 \text{ in.}$$

Now utilizing the formula for the thread stripping area listed with Tables 1 and 2 in Chapter 7, we get:

$$AS_s = \frac{\pi L_e K_{nmax}}{p} [0.5p + 0.57735(E_{smin} - K_{nmax})]$$

$$= \frac{\pi \times 0.75 \times 0.6418}{0.1} [0.05 + 0.57735(0.6851 - 0.6418)]$$

$$= 1.134 \text{ in.}^2$$

Design Problem 4

Calculate the tensile stress and thread root areas for the UNC, inch series thread having the basic geometry and dimensions equal to those given in Design Problem 3.

Solution: Using the formulas for AS and AR listed with Tables 1 and 2 of Chapter 7, we have:

$$AS = 0.7854 \left(D - \frac{0.9743}{n}\right)^2$$

$$= 0.7854(0.75 - 0.09743)^2$$

$$= 0.334 \text{ in.}^2$$

and

$$AR = 0.7854 \left(D - \frac{1.3}{n}\right)^2$$

$$= 0.7854(0.75 - 0.13)^2$$

$$= 0.302 \text{ in.}^2$$

Design Problem 5

Estimate the proof load for 1-1/4 inch diameter carbon steel fastener in the grade designation of A490, having 7 threads per inch.

Solution: The basic tensile stress area is:

$$AS = 0.7854 \left(D - \frac{0.9743}{n}\right)^2$$

$$= 0.7854 \left(1.25 - \frac{0.9743}{7}\right)^2$$

$$= 0.9691 \text{ in.}^2$$

From Table 3, the relevant proof stress is 120 ksi. Hence, the required proof load becomes:

$$0.9691 \times 120 = 116.3 \text{ kips}$$

Note that Table 4 for this condition recommends a round number of 116 kips.

Design Problem 6

It is desired to select a carbon steel, externally threaded metric fastener in the property class of 4.6 to carry a maximum load of 98 kilonewtons. Consult the appropriate tables to determine the nominal fastener diameter and the thread pitch.

Solution: From Table 7 the product size can be obtained directly as M20 × 2.5. The corresponding stress area from Table 2 is 245 mm². This is consistent with the tensile strength given in Table 6 as 400 MPa.

Since 1 MPa is equal to one newton per square millimeter, we get:

$$400 \ (\text{N/mm}^2) \times 245 \ \text{mm}^2 = 98,000 \ \text{N} = 98 \ \text{kN}.$$

Design Problem 7

The effective lengths of a machined bolt shank are 1.5 and 2.0 inches corresponding to 0.75 and 0.625 inch diameters, respectively. Calculate the total change in length of this bolt under axial load of 10,000 lbs. The material is aluminum. Check the nominal stress at the minimum diameter.

Solution: The appropriate formula for this case follows directly from Eq. (1).

$$\Delta L = \frac{4F}{\pi E} \left(\frac{L_1}{D_1^2} + \frac{L_2}{D_2^2} \right)$$

$$\Delta L = \frac{4 \times 10,000}{\pi \times 10 \times 10^6} \left(\frac{1.5}{0.75^2} + \frac{2}{0.625^2} \right)$$

$$= 9.9 \times 10^{-3} \ \text{in}.$$

The tensile stress is found from Eq. (4).

$$S_t = \frac{4F}{\pi D_2^2}$$

$$= \frac{4 \times 10,000}{\pi \times 0.625^2}$$

$$= 32,595 \ \text{psi}.$$

Note that this is a significant stress value for the material assumed and the fastener structural integrity may have to be evaluated with special regard to geometrical discontinuities and other design constraints.

Design Problem 8

The axial force on 1-1/4 inch fastener under torqued conditions was measured to be 50,000 lbs. The number of threads per inch for the coarse thread
series is 7. Assuming international thread profile and the coefficient of friction equal to 0.2, calculate the external torsional moment required to twist
the body of the fastener.

Solution: According to the American practice pitch is calculated from the
number of threads per inch as follows:

$$p = \frac{1}{n}$$

so that

$$p = \frac{1}{7} = 0.1429 \text{ in.}$$

The half-angle of the international thread profile is $\beta = 30°$. Hence, the
appropriate design formula for this problem can be obtained from Eq. (7).

$$M_{to} = F'(0.023 + 0.577\mu D)$$

Hence substituting the relevant numerical values gives:

$$M_{to} = 50,000(0.023 + 0.577 \times 0.2 \times 1.25)$$

$$\cong 8370 \text{ lb-in.}$$

Design Problem 9

Misalignment of flange surfaces is such that the fasteners are bent to the
radius of curvature equal to about 200 in. Assuming the fastener material
to be steel, calculate the external bending moment and the corresponding
bending stress for a fastener diameter of 0.875 inches.

Solution: Since the modulus of elasticity is 30×10^6 psi, Eqs. (9) and
(10) give:

$$M_b = \frac{\pi \times 30 \times 10^6 \times 0.875^4}{64 \times 200} = 4315 \text{ lb-in.}$$

and

$$S_b = \frac{30 \times 10^6 \times 0.875}{2 \times 200} = 65{,}630 \text{ psi}$$

Design Problem 10

Consider the special joint where an aluminum alloy fastener of the 2024-T4 type carries an external load in shear as shown in Fig. 3. Assuming 3/8 in. UNC fastener, calculate the maximum shear load this joint can carry.

Solution: For the nominal size of 3/8, Table 1 indicates 16 threads per inch for the UNC class of this product. Also Table 16 gives the maximum available shear strength as 37,000 psi for the 2024-T4 alloy.

Utilizing, now, Eq. (6), we obtain:

$$Q_{max} = \left(\frac{\pi \times 0.375^2}{2} - \frac{1.35 \times 0.375}{16} + \frac{0.75}{256} \right) \times 37{,}000$$

$$= 7111 \text{ lb}$$

Note that such a high load capacity is achieved because the fastener works in double shear.

Design Problem 11

When the amount of bolt preload is specified the required initial torque can be determined with good accuracy from Eq. (12). Assuming that the helix angle can be expressed in terms of the thread pitch and making the mean radius of thread approximately equal to nominal fastener radius, develop a simplified theoretical formula for this case. Assume also the half-angle of thread profile to be 30°, and $R_0/r \cong 1.4$.

Solution:

$$\tan \alpha = \frac{p}{\pi D}$$

$$R_0/r = 1.4$$

$$r = \frac{D}{2}$$

$$\cos \beta = 0.8660$$

Hence the revised formula becomes:

$$M_t = F_i D \frac{1.28 \mu D + 0.16 p - 0.26 p \mu^2}{D - 0.36 p \mu}$$

Design Problem 12
Show that the conventional formula for bolt preload, often quoted in machine design practice, can be obtained directly from Eq. (12) given in this book if we assume the following geometrical and frictional characteristics.

Half angle of thread profile, $\quad \beta = 30°$

Mean radius of thread, $\qquad r \cong \dfrac{D}{2}$

Helix angle of thread, $\qquad \alpha \cong 0$
Coefficient of friction, $\qquad \mu = 0.16$
Radius of friction area, $\qquad R_0 = 1.4r$

Solution:

$\cos\beta = 0.8660$

$\tan\alpha = 0$

Hence, Eq. (12) gives:

$$M_t = F_i \frac{D}{2}\left(\frac{0.16}{0.866} + 1.4 \times 0.16\right)$$

or

$$F_i \cong 5\,\frac{M_t}{D}$$

Note that this is Eq. (17). The value of K factor in this case is 0.2 which corresponds to the "as received" condition of a steel fastener as indicated in Table 19. Some of the "lubricated" values of μ are given in Table 20. Hence we see that Eq. (17) is generally conservative and safe to use in the determination of the tightening torque M_t.

Design Problem 13
Steel fastener is used in a mechanical joint using metal-to-metal contact where the overall stiffness of the joint is $K_c = 35 \times 10^6$ lb/in. Assuming that the fastener holding this connection has nominal diameter $D = 0.5$ in. and is 2 inches long calculate the percentage of the external load F_e which should be added to the fastener preload.

Solution: The formula for calculating the spring constant of the fastener is obtained from Eqs. (3) and (5).

$K_f = 0.7854 \; ED^2/L$

This gives:

$K_f = 0.7854 \times 30 \times 10^6 \times 0.5^2/2$

$\quad = 2.95 \times 10^6 \; \text{lb/in.}$

The required percentage formula follows from Eq. (19).

$$C_\% = \frac{100 K_f}{K_f + K_c}$$

Hence we have:

$$C_\% = \frac{100 \times 2.95}{2.95 + 35} = 7.8$$

Therefore 7.8% of the external load F_o should be added to the preload in order to estimate the actual load carried by the steel fastener used in this design problem.

Design Problem 14
The original preload of the fastener in a mechanical connection is given as $F_i = 50,000$ lbs. The design ratio of the fastener to joint stiffness is known to be 0.2. Assuming that the joint is expected to carry the external load of 40,000 lbs, estimate the actual load carried by the fastener.
Solution:

$$\frac{K_f}{K_c} = 0.2$$

$K_f = 0.2 \; K_c$

Substituting in Eq. (19) gives:

$$C = \frac{0.2 K_c}{0.2 K_c + K_c} = 0.167$$

Hence, using Eq. (18) gives:

F = 50,000 + 0.167 × 40,000

= 56,670 lbs

Design Problem 15
A soft copper gasket is used with a slender bolt to carry an external load of 20,000 lbs. If the bolt is initially preloaded to 40,000 lbs calculate the actual load in the bolt.

Solution: The stiffness parameter C for this case can be directly obtained from Table 20, as C = 0.50. Hence using Eq. (18) gives:

F = 40,000 + 0.5 × 20,000

= 50,000 lbs

Design Problem 16
The actual load monitored on a fastener was found to be equal to 55,000 lbs. Assuming that the preload and the external load on the joint were 50,000 lbs and 60,000 lbs, respectively, calculate the overall stiffness coefficient of the joint, and the ratio of the two spring constants.

Solution:
From Eq. (18):

$$C = \frac{F - F_i}{F_e}$$

$$= \frac{55,000 - 50,000}{60,000}$$

$$= \frac{1}{12}$$

From Eq. (19):

$$\frac{K_f}{K_c} = \frac{C}{1 - C}$$

$$= \frac{1/12}{1-(1/12)}$$

$$= \frac{1}{11} = 0.091$$

Design Problem 17

A mechanical joint held by fasteners contains a gasket of stiffness K_g placed between flanges of stiffness K_j. Assuming that $C = 1/12$, as shown in Design Problem 16, calculate the ratio of fastener to flange constant if $K_g/K_j = 1/5$.

Solution: Divide numerator and denominator of Eq. (23) by $(K_j)^2$.

$$C = \frac{[(K_f/K_J)] [(K_g/K_j + 1)]}{[(K_f/K_j) + (K_f/K_j)(K_g/K_j) + (K_g/K_j)]}$$

Substituting the numerical values gives:

$$\frac{1}{12} = \frac{(K_f/K_j)[(1/5) + 1]}{[(K_f/K_j) + (1/5)(K_f/K_j) + (1/5)]}$$

or

$$\frac{K_f}{K_j} = 0.015$$

Design Problem 18

Assuming that the spring constant of the joint is ten times the fastener stiffness, calculate the total fastener load if the initial preload is 20,000 lbs. The external load on the joint is 12,000 lbs.

Solution: From Eq. (19), using $K_c = 10K_f$

$$C = \frac{1}{11}$$

Hence Eq. (18) gives directly:

$$F = 20,000 + \frac{12,000}{11}$$

$$= 21,000 \text{ lbs}$$

39

Comments on Quality Control of Fastener

The fundamentals of dealing with fastener selection cannot be complete without some observations on quality control procedures which, to no surprise, are a hot topic in the fastener world of today. There are many examples, even in modern industries, where as-received fasteners are not always what they are claimed to be. And it appears that the general situation is not likely to get very much better.

By a strict definition, quality control represents those quality assurance actions which provide a means to monitor and measure the characteristics of an item, material or process, in relation to the established requirements. In the case of a threaded fastener for instance the following three major elements may be of special importance:

1. Physical size and material
2. Environmental constraints
3. Service conditions

The foregoing three elements cover various engineering aspects related to the selection and installation procedures for fasteners which include joint design, fabrication, preload, and assembly. Several design and preload characteristics have been outlined throughout this book with the aim of helping the

reader to make a proper materials choice and to assess some of the dimensional and strength properties of a given fastener. Existing standards are
based essentially on hardness specifications whenever there is a viable correlation between the hardness and strength properties of a metallic fastener. One
can argue the point that practical data, including such a correlation, may not
be readily available for a particular application. The alternative would be to
resort to various other verification methods which are for the most part designed for the direct control of preload. The application of direct torque
methods, of course, may not be very accurate if only 10% of the total
wrenching torque can be utilized during the bolt (fastener) tensioning.

Before addressing some of the shortcomings and dangers existing in
modern fastener applications, involving materials or processing, let us look
at the quality control measures where hardness testing is not specified. We
will try to indicate the potential accuracy, limitations, and, whenever possible, economic considerations. This last observation is extremely important
because of recent trends to extensive litigation and increased safety awareness. There is now a renewed interest in optimization where economics and
risk assessment variables begin to play a special role. Of these two, the risk
assessment variable is perhaps the most difficult to define, control, and
accept as a design criterion. It also appears to be equally difficult to accept
as a management tool.

The strain gage applied to the bolt for an exact reading of the strain
assures the best accuracy of ±2%. This is certainly a direct method of judging
the preload. Unfortunately the cost is rather high, amounting to at least
$200 per bolt and a minimum of $1,000 in capital cost for a readout. Its
application then is pretty much limited to larger bolts and the method is
sensitive to misalignment.

When an active strain gage washer is applied under a standard nut or bolt
one can continuously monitor the preload provided the bolt is relatively long.
The main disadvantage of this approach however is cost. In terms of current rates, we may be looking at $500 per bolt plus the expense of the
readout equipment.

Where the accuracy of measuring equipment does not have to be better
than ±15%, depress-dimple washers can well be used. Such washers have
deformable protrusions on one surface and are usually installed like regular
washers. As the bolt is tightened a feeler gage is used between the washer
and the bolt to determine the degree of preload. These washers are not reusable and require extra bolt length. Similar accuracy of readout can also
be obtained with the so-called dual-washers or spinning-ring washers.

In the former case two concentric washers are installed where one is slightly higher than the other. The shorter washer is generally free to turn until the specified torque is reached. In the case of a spinning-ring washer, the hardware is reusable but there is some danger that the system can be overtorqued. This last method, therefore, cannot be recommended.

One of the more sophisticated control methods depends on ultrasonic length measurement. While there are a number of advantages in using this system the required instrumentation is expensive. Similarly using tension control through the application of electronics, good accuracy can be assured with automatic control of bolt quality. When the proper preload is attained, the torque wrench can be shut off. The final torque and the angle of twist can then be compared with standard values allowing detection of improper bolt grade, thread flaws, bottoming-out, or a catastrophic bolt failure. Unfortunately again, the test equipment is quite expensive and at least on the order of $15,000.

The conventional use of a torque wrench, which is properly calibrated, is quite economical and acceptable provided less accurate measurements can be tolerated. This method is rather insensitive to misalignment and applies to a wide range of bolt sizes. A derivative of this approach involves the so-called turn-of-nut method. The nut is first snugged up and then rotated through a specified angle to reach the prescribed preload. Advantages include low cost and no special limitations on the size of bolt tested. However very careful calibration is required and the method is sensitive to misalignment and variations in thread length.

For special applications hydraulic tensioning is recommended which can give about ±10% accuracy. The procedure here is that the bolt is held by a hydraulic tensioning device at a prescribed load. Subsequently the nut is snugged into place and the hydraulic pressure released. The approximate cost of tooling and application hardware for this case is about $2,000.

For an engineer dealing with only a few fasteners or for a small manufacturer trying to set up a quick test of as-received fasteners, either a hardness test or the depressed-dimple washer method may fill the general need for reasonable accuracy and economy. The question still remains, of course, how extensively the test test program should be developed in order to qualify the product. If we talk about larger orders of fasteners running into many thousands, the existing specifications and standards fall short of preventing some of the bad fasteners from penetrating the so-called "within spec envelope." Certain companies, particularly in the automotive field, use literally tons of fasteners so that even a fraction of a cent per fastener can add up to significant savings. The cost of quality control therefore should enter the optimization equation in the competitive world, including domestic and foreign markets. The decision on the number of bolts, for instance, to be tested for qualifying a given order is the most difficult and speculative. We ask ourselves what should be the statistically admissible sampling program in order not to just stay in the "within spec envelope" but to reduce the level of a potential disaster if the supposedly "in spec" fasteners fail. Recent experience shows that fasteners can fail although their hardness (and therefore strength) are still found to be within a legitimate specification. In order to bring this argument home let us pause a moment to recall a few publicized cases. Here the prognosis for the future does not look much better even for such mundane and simple machine and structural elements as threaded fasteners.

Take for instance the case of a known aircraft manufacturer who had to replace thousands of high-strength bolts following some fatal and near-fatal accidents. It seems that no method or number of tests, on an "as-received" basis, could have anticipated the problem that these bolts would corrode and crack after a specific length of service which damaged their cadmium coating. The layer of cadmium was originally intended to act as a protective shield. What kind of quality control procedures should this particular manufacturer have instituted in order to avoid a series of these problems?

Let us look at other examples where high-strength bolts were involved. About six years ago it became necessary to replace some 12,000 bolts in the roof of a large fieldhouse in the U.S. Eventually it became rather obvious, during the planning and construction phase, that high-strength steel bolts should have never been used, even with the best certification procedures. About the same time certain faulty high-strength steel bolts were discovered on the assembly line for non-commercial aircraft. This discovery delayed the production of the sophisticated plane until a search removed all the faulty joints, some of them from already completed planes. There is little doubt that undetected bad fasteners could have resulted in serious consequences. About five years ago nearly six million motor vehicles were called back to the factory in order to replace the high-strength steel bolts supporting the rear suspensions of various models built between 1978 and 1981. The basic problem was that standard tests could not have reliably predicted the susceptibility of these bolts to stress corrosion and cracking.

The well informed engineer is now sufficiently warned about the shortcomings and dangers of misusing high-strength materials and existing inspection standards in the area of fasteners. In general, however, specified inspection lot sizes are too small to be meaningful. For instance, if you manufacture 300,000 A325 bolts, you may have to test only 13 bolts for hardness in order to satisfy a particular specification. Furthermore, since a number of standards are based on the minimum hardness only, many as-received A325 bolts can technically be of the A490 class if the manufacturer over-hardens the A325 bolts. The impression that we get a better product for the money because of the increased strength can eventually hurt us since higher strength means greater susceptibility to stress corrosion and fatigue failure. It all boils down to the fact that the user should develop his own quality control tests and procedures of risk assessment analysis for all critical applications of fasteners.

The promotion of light-weight construction and efficiency in various industries since the early 1970s has resulted in the greater use of high-strength steel. At present about five million tons of high-strength steel are produced in this country annually. As this material came to be used in many sophisticated designs the tolerance band for even the smallest fabrication error has become very narrow. This problem has now intensified because the overall quality of high-strength steel in recent times has not improved significantly. This is the basic materials and processing shortcoming which demands extensive field testing and more elaborate quality control procedures.

In closing this brief encounter with the quality assurance issues relevant to fastener applications it should be stated that currently recognized inspection standards should be carefully reviewed and modified to keep pace with developments in safety awareness and litigation practices. Designers, and the companies they represent, should stand on their own feet in scrutinizing as-received fasteners. New information on material properties, performance, and special problem areas will become even less available as the litigation procedures create additional layers of technical secrecy.

The foregoing comments on quality control, specific cost figures, and applications of threaded fasteners, may or may not be pertinent to future economic and technological developments. The basic requirements of quality and safety assurance however are difficult to dispute.

40

Basic Units

The International System of Units (SI) has evolved over the past several years. It is essentially built upon three base units, known as meter, kilogram, and second.

As far as the mechanical designer and hardware manufacturer are concerned, the most immediate choice of units appears to be stress, elastic constants, and physical dimensions. The SI system differentiates rather strongly between the concepts of weight and mass. Weight is defined as the force of gravity, and it is expressed in newtons. The newton, which is then a unit of force, is that magnitude of a force which gives to a mass of 1 kilogram an acceleration of 1 meter per second per second ($1m/sec^2$). On this basis, the stress, defined as the force divided by the area, would be expressed as newtons divided by meters squared. In the SI system 1 pascal (Pa) is equal to 1 newton per square meter, which is the formal SI unit for pressure, stress, material strength or elastic constants. The conventional acceleration of gravity, g, is now 9.8 m/sec^2.

For straightforward applications in fastener engineering the proposed unit "pascal" is not only too small for practical purposes, but it is also quite superfluous. It would appear much more reasonable to use newtons per area instead of pascals. This author prefers newtons per square millimeter and there are several practical reasons why the use of N/mm² should be promoted.

In the first place, the dimensions of countless machine components and structural elements, as well as fasteners, found worldwide, are expressed in millimeters. Secondly, the absolute magnitude of the newton force appears to be compatible with the area of a square millimeter and in such a case 1 atmosphere, for instance, is approximately equal to 0.1 N/mm². Using these units the modulus of elasticity of steel will work out to be 2×10^5 N/mm² and the approximate yield strength of a common structural steel could be remembered as 250 N/mm². Numerically, of course, 1 megapascal is equal to 1 newton per square millimeter, as shown by the following transformation:

$$\frac{N}{m^2} = \frac{N}{(m \times 10^{-3})^2} = \frac{N}{m^2 \times 10^{-6}} = \frac{10^6 N}{m^2} = 1 \text{ MPa}$$

The use of N/mm² should automatically correct our present lack of uniformity caused by the open choice of pascal, kilopascal, megapascal, and other potential multiples of this small and cumbersome unit.

Although the English system of units predominates in this book, the more commonly used conversions are given in Table 22.

Table 22 Basic Conversions

1 lb	=	4.4482 N
1 kg	=	9.8066 N
newton (N)	=	unit of force
1 in.	=	25.4 mm
1 m.	=	39.37 in.
1 lb-in.	=	112.9842 N-mm
1 lb/in.	=	0.1751 N/mm
1 psi	=	6895 Pa
pascal (Pa)	=	unit of pressure
Pa	=	N/m²
1 MPa	=	145 psi
1 atmosphere	=	0.1 N/mm²
1 psi	=	0.006895 N/mm²

The values given in Table 22 have been rounded off slightly for practical reasons. The great majority of design formulas are homogeneous, so that a consistent set of units can be employed directly. The mechanics of a typical conversion will be now illustrated. Suppose we have a formula for a shear stress due to a twist of bolt shank

$$\tau = \frac{M_t r}{I_p}$$

where

M_t = twisting moment, lb-in.

r = radius of shank, in.

I_p = polar moment of inertia, in.^{4}

Hence, assuming unit values, we get

$$\tau = \frac{4.4482 \text{ N} \times 0.0254 \text{ m} \times 0.0254 \text{ m}}{(0.0254 \text{ m})^4} = 6895 \text{ Pa}$$

Suppose we have

M_t = 500 lb-in = 500 $\times$ 4.4482N $\times$ 0.0254 m = 56.49 N-m

r = 2 in. = 2 $\times$ 0.0254 m = 0.0508 m

I_p = 0.4in.^{4} = 0.4 $(0.0254)^4$ m^4 = 1.6649 $\times$ 10^{-7} m^4

Then using the English system

$$\tau = \frac{500 \times 2}{0.4} = 2{,}500 \text{ psi}$$

and the SI system gives

$$\tau = \frac{500 \times 112.9842(\text{N} - \text{mm}) \times 2 \times 25.4 \text{ (mm)}}{0.4 \times 25.4^4 \text{ (mm)}^4} = 17.24 \text{ N/mm}^2$$

Note that 2,500/17.24 = 145. This is the conversion factor between MPa and N/mm^2, as shown in Table 22.

It appears that the metric unit of stress defined as N/mm^2 is rather convenient. It also reminds us of the old metric system, practiced in many corners of the world, where the measure of pressure, stress, or a mechanical property was kg/mm^2.

Index

195

Printed in the United States
by Baker & Taylor Publisher Services